T0220452

Fundamentals of Atomic Force Microscopy

Part I: Foundations

Lessons from Nanoscience: A Lecture Note Series

ISSN: 2301-3354

Series Editors: Mark Lundstrom and Supriyo Datta
(Purdue University, USA)

"Lessons from Nanoscience" aims to present new viewpoints that help understand, integrate, and apply recent developments in nanoscience while also using them to re-think old and familiar subjects. Some of these viewpoints may not yet be in final form, but we hope this series will provide a forum for them to evolve and develop into the textbooks of tomorrow that train and guide our students and young researchers as they turn nanoscience into nanotechnology. To help communicate across disciplines, the series aims to be accessible to anyone with a bachelor's degree in science or engineering.

More information on the series as well as additional resources for each volume can be found at: http://nanohub.org/topics/LessonsfromNanoscience

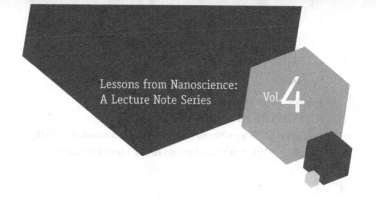

Lessons from Nanoscience:
A Lecture Note Series

Vol. 4

Fundamentals of Atomic Force Microscopy

Part I: Foundations

Ronald Reifenberger

Purdue University, USA

World Scientific

NEW JERSEY · LONDON · SINGAPORE · BEIJING · SHANGHAI · HONG KONG · TAIPEI · CHENNAI · TOKYO

Published by

World Scientific Publishing Co. Pte. Ltd.

5 Toh Tuck Link, Singapore 596224

USA office: 27 Warren Street, Suite 401-402, Hackensack, NJ 07601

UK office: 57 Shelton Street, Covent Garden, London WC2H 9HE

British Library Cataloguing-in-Publication Data

A catalogue record for this book is available from the British Library.

Lessons from Nanoscience: A Lecture Note Series — Vol. 4
FUNDAMENTALS OF ATOMIC FORCE MICROSCOPY
Part I: Foundations

ISBN   978-981-4630-34-4
ISBN   978-981-4630-35-1 (pbk)

In-house Editor: Song Yu

Typeset by Stallion Press
Email: enquiries@stallionpress.com

Printed in Singapore

Preface

Since dedicated courses are unlikely to spring up in university curricula to teach the fundamentals of Atomic Force Microscopy (AFM), it is useful to have a set of notes that interested students can work through in an independent, self-study mode at their own pace, without the benefit of a rigorous class schedule.

The lecture notes *Fundamentals of Atomic Force Microscopy, Part I: Foundations* were written to address this issue. The notes have slowly grown from three one-semester courses given at the graduate level at Purdue University in 2009, 2010, and an online NanoHUB-U course in 2012. In preparing and teaching these courses, I have all too often realized first-hand that students lack the background required to read the AFM literature critically. They often do not know the vocabulary, the core concepts, or the mathematics required to understand the fundamental principles underlying the operation of an AFM.

Why make an effort to work through these notes?

> First and foremost, an interdisciplinary background is required to meaningfully attack the truly challenging problems facing us today. The lecture notes attempt to fill many of the gaps in knowledge that traditional students (i.e., students who follow a traditional discipline-based course of study) often have when embarking on interdisciplinary studies.
>
> Secondly, the lecture notes are designed to be read in parallel with the two-part nanoHUB-U online course called *Fundamentals of AFM; Parts I and II* that was developed during the summer of 2012. The video lectures are freely available over the internet (see https://nanohub.org/courses/AFM1 and https://nanohub.org/courses/AFM2). While the notes are not

completely synchronized with the nanoHUB-U lectures, there is a sufficiently strong overlap that the interested student should benefit by following both the notes and lectures in parallel.

The pace of the online lectures is formidable and inevitably, typos and missing factors of π creep in. Most of these errors are corrected in the lecture notes and hopefully new errors have not reappeared during the process. If they have, I would be most happy if the reader takes the time to send me an email identifying the issue and listing the page or equation along with an explanation of the problem.

At the beginning of each chapter, I make an effort to identify why and how the chapter content is necessary for the intelligent use of an AFM. In each chapter, I develop topics that should be helpful to both acquire meaningful AFM data and to analyze it properly. I also attempt to identify the traditional academic disciplines that a four-year undergraduate might need to better appreciate the chapter content.

Lastly, my colleague Arvind Raman is preparing a set of lecture notes to cover Part II of the video lecture series, which focusses on dynamic AFM techniques and methods.

The notes are *not* intended to be exhaustive and they touch only on what I judge to be of fundamental interest. The AFM expert will not learn anything new since the notes are written with the beginning students in mind. I have attempted to provide specific references to the original literature whenever possible, but I have adopted the approach that in lecture notes, it is perhaps the ideas rather than the correct attribution that is most important.

R. Reifenberger
W. Lafayette, Indiana
April 3, 2014

...to the Student

There are many reasons why you are reading these lecture notes. Perhaps you have tried to perform an AFM experiment, you have obtained some results, but you don't know how to interpret the data. Or perhaps you are taking a short course on AFMs required by your plan of study. Or maybe you are just interested in learning more about the AFM technique in general. For whatever reason, we hope you will know more about how AFMs function after working through these notes.

The lecture notes are an outgrowth of previous semester-long courses offered at Purdue University in 2009, 2010 and 2012. The latest course in 2012 was incorporated into a nanoHUB-U set of online lectures that have been widely viewed on the web (see https://nanohub.org/courses/AFM1 and https://nanohub.org/courses/AFM2). Many of the students taking these online courses requested a set of written lecture notes. These lecture notes are a direct outgrowth of their requests.

While developing the *Fundamentals of AFM* lectures, I have come to appreciate a number of issues. First, the proper use of an AFM truly requires in-depth knowledge from many different traditional fields of study. Second, the ease that students integrate the many different topics discussed depends to some extent on their prior academic training. Third, there seems to be an overwhelming tendency for qualitative, hand-waving arguments to interpret experimental AFM results.

The lecture notes are organized with these issues in mind. Since not all students have the required background to understand the wide variety of topics discussed, I have attempted to summarize information relevant to AFM in one place. Since understanding tends to follow prior course work, I have provided both qualitative and quantitative background information supported by mathematical derivations for the broader audience that may read these notes. While some may think a certain analysis is trivial, others may appreciate it because they happen to lack the necessary background that it discusses.

Finally, I include a number of worked-out examples to illustrate the ideas that have been developed. Only by attempting to calculate various quantities do you truly test your limits of understanding. To further this goal, I have made an attempt to standardize the notation found in the literature and use consistent symbols and formalism throughout the lecture notes. Homework problems are provided at the end of each chapter. References to the original literature are also provided as necessary.

Another feature of these lecture notes is the use of the free online AFM software VEDA developed by A. Raman and his students. Worked examples, which use the VEDA code to answer questions that cannot be solved analytically, are included. Knowing how to use VEDA will allow you to understand AFM data quantitatively and to devise better AFM experiments as you pursue your research interests.

Lastly, these notes have been prepared for students, not for AFM experts. There are already many reviews of AFM written by experts for experts. Unfortunately, the material contained in these reviews is often not readily accessible to the beginning student without a prior discussion of background material. For this reason, many of the discussions found in these lecture notes are intentionally pedagogic in-nature to better convey an understanding of the important fundamentals underlying scanning probe microscopy.

I would greatly appreciate knowing about my missteps, mistakes and oversights that I have made along the way. It's the only way an exposition of this material will improve.

R. Reifenberger
W. Lafayette, Indiana
May 3, 2014

Acknowledgments

While few who read this book will read this page, it is nonetheless important to thank those who have contributed to these lecture notes.

First and foremost are the ~50 Purdue students who have satisfied MSc and PhD requirements working in my lab since I first became intrigued by Binnig's controllable tunnel gap in 1983. For ~30 years we have designed, built and used scanning probe microscopes to investigate the amazing nanoscale properties of matter. These lecture notes are in no small part an attempt to answer the collective questions my students have asked me throughout the years.

The long term collaboration with Prof. Arturo Baro's *Laboratorio de Nuevas Microscopías* at the Universidad Autonoma de Madrid has contributed immensely to my understanding of scanning probe microscopy. The many discussions with Julio Gomez-Herrero as we walked through Madrid during the early days after the invention of the STM are especially noteworthy.

The decade long collaboration with my colleague from Purdue, Arvind Raman, has been instrumental in my understanding of cantilever dynamics and critical for the *Fundamentals of AFM* lectures that are now available on the web. His insistence on mathematical clarity and rigor has been a source of inspiration to an experimentalist who often relies on intuition, simple models and back of the envelope estimates for guidance on how best to proceed.

These notes have benefitted greatly from critical comments and suggestions by Maria Jose Cadena, Ryan Wagner and Scott Crittenden. Scott's yeoman-like attention to detail is especially notable for improving the content and presentation of the material found in these notes.

Lastly the life-long support of my wife Ellen cannot go without mention. Over the years, she has endured countless delayed meals, innumerable late-night trips back to the lab, and postponed many a weekend excursion with a grace and presence that few could understand much less realize. Perhaps now that these lecture notes are complete, I can accommodate her dream to spend a summer touring the Canadian Maritime Provinces.

R. Reifenberger
W. Lafayette, IN
May 25, 2014

Contents

Chapter 1

Introduction to Scanning Probe Microscopy

Today's research laboratory is required to solve difficult problems that span multiple disciplines. Advanced techniques are required to answer pressing questions related to adhesion, bonding, contamination and surface cleanliness, corrosion, surface morphology, surface roughness, surface topography, failure analysis, process monitoring, surface chemistry, biological characterization, local surface properties — both electrical and mechanical, and thin film analysis. Rarely can one analytical technique effectively span such a wide range of applications. The rapid rise of scanning probe microscopy (SPM) provides a truly marvelous tool that provides useful information about all these topics and many more.

Few scientific instruments have received as much attention and enjoyed such rapid growth as the atomic force microscope (AFM). The inherent simplicity of the AFM coupled with its ability to apply nanoNewton forces to surfaces with sub-nanometer lateral precision have led to a significant expansion in both the scope and context of the instrument. Originally used to probe the atomic roughness of a surface, the AFM has quickly evolved into a probe of surface forces using primarily only two modes of operation (contact and dynamic). Of particular significance are the rapidly evolving techniques that allow quantitative material property maps of surfaces with nanometer-scale resolution. Furthermore, the ability to position and precisely move a biased tip has been exploited to demonstrate novel nanoscale device fabrication. Taken together, these developments have led to the widespread use of AFM in all fields of science and engineering.

The intelligent use of SPM and AFM requires broad training in a multitude of different disciplines spanning many areas of science and engineering.

New graduate students, when asked to use an AFM ask many questions: How do I choose a cantilever? How fast should I scan? How do I optimize the feedback? What should I do to reduce noise? How can I improve resolution? What's the best way to prepare a sample? After many years of answering such "how to?", "what should I do?" questions one at a time, I found it more useful to first teach my students at Purdue the fundamentals of how an AFM works. These lecture notes are the logical consequence of this approach. In writing this book, I have two main goals: (i) to convey to a beginning student the scope of knowledge required to properly use an AFM, and (ii) to convey a clear understanding of the physics and mathematical models underlying AFM. If these two goals are met, then students should have the necessary tools to provide answers to many important and vexing questions as they arise. There are numerous seminars, power point presentations and monographs available on the web that survey different imaging modes and provide helpful hints to technical questions about image optimization and sample preparation. What is lacking is an extended discussion about the physics and mathematics underlying the AFM coupled with a discussion about the fundamentals of AFM design, its operation and its use. We find very little in the way of a systematic discussion of the fundamentals of AFM, a topic that forms the focus of these lecture notes.

1.1   Historical Perspective

The generic SPM is an extremely versatile instrument that has steadily evolved from its invention in the early 1980s. In these lecture notes, SPM is used to broadly denote the two most popular scanning probe instruments, the scanning tunnelling microscope (STM) and the atomic force microscope (AFM). SPMs are now routinely available in many research labs throughout the world and are widely heralded for ushering in the study of matter at the nanoscale.

The underlying principles of an SPM are quite simple but yet completely different in many significant ways from more traditional microscopes. Essentially, the SPM works by positioning a sharp tip (often called a proximal probe) about 1 nanometer above a substrate. The highly local information provided by the microscope is achieved by a combination of the sharpness of the tip as well as the small separation between the tip and substrate. The critical feature of any SPM is the ability to maintain a constant tip–substrate distance (with a precision approaching a few picometers) while the tip is rastered across the substrate in a highly controlled way. While it is important that the tip–substrate distance be held constant, it is surprisingly difficult to accurately know the exact value of this distance.

To achieve high precision, a signal must be acquired that is very sensitive to the tip–substrate separation. The exact physical origin of this signal then determines the property of the substrate that is mapped. A key discovery during the development of SPMs was the realization that with a sufficiently sharp tip, a quantitative 3-dimensional image of surfaces can be obtained, often with atomic resolution.

The worldwide interest in scanning probe instruments was ignited by the research accomplishments of G. Binnig and H. Rohrer at the IBM Zurich Research labs in Switzerland [binnig87]. These two individuals shared the 1986 Noble Prize in Physics (along with E. Ruska, inventor of the electron microscope) for their seminal work in the invention of SPM [binnig82], [binnig83a], [binnig83b], [binnig86]. A reading of the published literature has revealed relevant prior art that resemble the implementation of the first SPM in the early 1980s. Work on surface profilers (using optical deflection techniques similar to those used in current scanning force microscopes) can be found in the published work of G. Shmalz that dates to 1929 [shmalz29]. R. Young, J. Ward and F. Scire developed in 1972 an instrument (called a topografiner) designed to measure the surface microtopography of a substrate [young71], [young72]. This work utilized a controllable metal-vacuum-metal separation to maintain a fixed tip–substrate distance, in some sense foreshadowing by some 10 years the tunnel gap approach developed independently by Binnig and Rohrer.

Before beginning an in-depth study of an AFM, it is useful to first discuss the general principles underlying *all* SPMs. The two widely-used families of SPMs — the STM for studying the surface topography of *electronically conducting* substrates and the AFM, developed to investigate the surface topography of *electrically insulating* substrates. The proliferation and development of SPM technology has greatly benefitted from parallel developments in both fields of STM and AFM, generating a wide variety of dual probe implementations of hybrid SPMs (sometimes called SxMs; where x stands for some physical variable of interest), which have led to simultaneous measurements with high lateral and vertical resolution not only of surface topography but also of other local properties of substrates.

1.2   The Need for a Scanning Probe Microscope

When invented, the STM was a unique instrument because it relied on proximal probe techniques to interrogate very local properties of an electrically conducting sample. The data obtained were able to resolve individual atoms by providing a 3-dimensional image, a seemingly commonplace occurrence today but quite a remarkable achievement in the 1980s.

The ability to view an inanimate object in 3-dimensions dates back to the 1840s when the stereo pair concept was invented, and has been used extensively for military and geographical (terrain) applications, and in more modern times for entertainment purposes. A map of individual atom positions was first achieved using the field-ion microscope [muller56] and had developed into an active field of research by the mid-1960s [muller65]. But the ability to couple these two capabilities into a single widely accessible instrument was a truly remarkable development that enabled world-wide experiments in many scientific disciplines.

Traditional electron microscopy (e.g., transmission electron microscopes (TEM) and scanning electron microscopes (SEM)) rely on the small size of the de Broglie wavelength of electrons to provide sufficient resolution to view sub-micron features on conducting surfaces. The use of these microscopes has grown continuously since their resolution surpassed optical microscopes shortly after their invention in the early 1930s. By the 1960s, TEMs were capable of 0.3 nm resolution while SEMs were able to resolve objects in the 15–20 nm range. By the 1980s, wide-spread improvements in instrumentation enabled analytical applications of electron microscopes that included energy dispersive x-ray spectra, electron energy loss spectroscopy, and the development of high resolution imaging theory [haguenau03].

In spite of these successes, the ability to combine compositional identification with electrical and/or mechanical measurements at select locations on a sample surface was a noticeable limitation of the many new surface science tools that were rapidly developed in the 1970s. This clearly identified need, evident in both academic and industrial research labs around the world, might explain the rapid acceptance of SPMs once their basic capabilities were demonstrated. While freely admitting that the attributes and shortcomings of each technique are a matter of personal taste, Table 1.1 attempts to compare TEM, SEM, and SPM.

Finally, it should be remarked that any student working in the world of sub-micron science or engineering requires a good knowledge of many

Table 1.1 A brief summary of the attributes of different sub-micron microscopies in common use in a modern research laboratory.

	TEM	SEM	SPM
Notable Attribute	Atomic Resolution	Depth of Field	High contrast in z plus high lateral resolution
Major Limitations	Extensive sample preparation of thin specimens; sample modification under e-beam irradiation; initial expense is high	Requires electrically conducting samples, extensive training often required, high cost of equipment maintenance	Speed of image acquisition, special training often useful
Dimensions Probed	2-D, planar	2-D, planar	3-D
Notable Capability	Atomic resolution	Chemical Composition	Topography plus Physical Properties
Environment	High Vacuum	Primarily Vacuum	Vacuum, Air, Liquid

characterization tools to produce credible research results. Knowing which tool to use in what situation can save considerable time in the pursuit of definitive answers to pressing problems. The choice of techniques are large and must include light microscopy, TEM, SEM, SPM, X-ray diffraction (XRD), X-ray Photoemission Spectroscopy (XPS) for elemental analysis, electron spectroscopy (UV photoemission, LEED) for surface analysis, and vibrational spectroscopy (HREELS, FTIR and Raman scattering) for molecular analysis.

1.3   The Scanning Tunneling Microscope

The STM was historically the first SPM and was introduced in 1982 by G. Binnig and H. Rohrer with the demonstration that a controllable vacuum tunnelling gap could be achieved between a sharp metallic tip and a conducting substrate [binnig82]. The vertical resolution of the STM is a few picometers while the lateral resolution can range down to $\sim$0.1 nm on an atomically flat substrate. STM images typically span an area ranging from a few nanometers to a few 100's of nanometers.

To understand tunnelling through a vacuum gap, knowledge of quantum mechanics, solutions to Schrödinger's wave equation and a basic understanding of electron states in metals are required [gomez05]. Since STM is

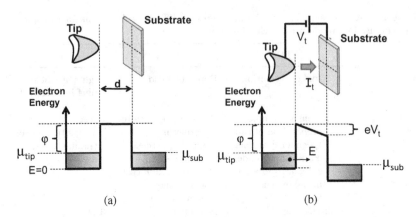

Fig. 1.1   In (a), a schematic of the potential barrier of width d (~1 nm) between a metallic tip and metallic substrate. In equilibrium, the electrochemical potential of the tip (μ_{tip}) and that of the substrate (μ_s) are aligned. The height of the potential barrier is $V_o = \varphi + \mu_{\text{tip}}$. In (b), the situation that develops when a bias voltage (V_t) is applied between tip and substrate. A tunnel current (I_t) comprised of electrons with various energies E can arise even though there is no physical contact between the tip and substrate.

not the main focus of this book, we refer those not familiar with these topics to textbooks that discuss tunneling in a straightforward way [tipler12]. Electron tunnelling is usually discussed for the ideal case when an electron with incident energy E encounters a barrier with a characteristic width d and a characteristic height V_o. The situation is shown in Fig. 1.1(a).

Such a barrier develops in the physical gap between a tip and substrate with a height $V_o = \varphi + \mu_t$ where φ (typically 4–5 eV; $1\,\text{eV} = 1.602 \times 10^{-19}$) represents the work function of the metal tip and μ_{tip} (typically 5–10 eV) represents the value of the Fermi energy, the most energetic electron in the tip. The presence of this barrier prevents the transit of a classical electron, but within the context of quantum mechanics, the electron is treated as a wave and it has a finite probability of penetrating the barrier.

The underlying physics required to understand how an STM operates begins by considering electrons incident upon a barrier at an energy $E < V_o$. Such electrons can quantum mechanically tunnel through the barrier with a transmission probability T that can be obtained from a time-independent solution to Schrödinger's Equation. For a rectangular barrier, the transmission probability is given by [tipler12]

$$T = \frac{4E(V_o - E)}{4E(V_o - E) + V_o^2 \sinh^2(\kappa d)} \quad E < V_o. \tag{1.1}$$

The wavevector of the electron when tunnelling is defined by

$$\kappa \equiv \frac{2\pi}{h}\sqrt{2m(V_o - E)}, \tag{1.2}$$

where m is the electron mass $(9.109 \times 10^{-31}\,\text{kg})$ and h is Planck's constant $(6.626 \times 10^{-34}\,\text{Js})$. When $\kappa d \gg 1$ (an approximation appropriate for STM experiments), Eq. (1.1) reduces to the well-known result that

$$T \simeq \frac{16E(V_o - E)}{V_o^2}e^{-2\kappa d}. \tag{1.3}$$

If $E \cong \mu_{\text{tip}}$ as is the case for low applied bias, then $V_o - E \cong \varphi$, the work function of the metal surface. Since φ is about $5\,\text{eV}$, the coefficient 2κ appearing in Eq. (1.3) is approximately $23\,\text{nm}^{-1}$.

For an electrical current I_t between the tip and substrate, a bias voltage V_t must be applied as shown in Fig. 1.1(b). This bias voltage will distort the shape of the square barrier, which will also be rounded and lowered in height by many-body electron correlation effects not considered here. The electric current between tip and substrate will be approximately proportional to the transmission probability defined in Eq. (1.3) for a square barrier. Since the measured current will be comprised of tunnelling electrons with different energies E, after integration over an appropriate range of energies, the tunnel current I_t will be given by an expression of the general form

$$I_t \simeq f(V_t, \varphi)e^{-2\kappa d}, \tag{1.4}$$

where $f(V_t, \varphi)$ is a function that depends on the applied voltage and the exact shape of the barrier under consideration. The exact barrier shape is difficult to determine, which explains why the analytical result for a square barrier is so often invoked. For applied voltage differences of $\sim 1\,\text{V}$, typical tunnel currents encountered in STM experiments lie between $0.01\,\text{nA}$ and $1\,\text{nA}$, depending on the value of d.

The strong exponential dependence of I_t with distance d is the important point to remember from this discussion. Rough estimates using Eq. (1.3) indicate that a change in the barrier width d by $0.1\,\text{nm}$ causes a change in I_t by roughly a factor of 10. This large amplification implies that small tip motions can be easily detected, measured and hence controlled. This key realization opens the door to a practical STM.

The exquisite sensitivity of I_t to tip–substrate separation d is used to monitor the vertical tip position above a substrate and hence transform tunnel current variations into high magnification images of a sample. Two modes of imaging have been developed: (a) constant height imaging in which

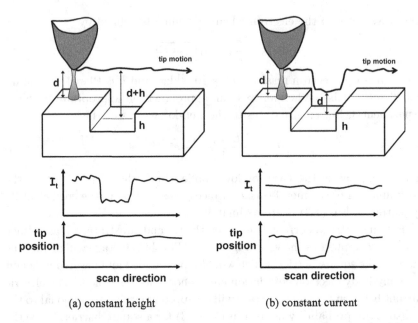

Fig. 1.2   A schematic illustrating two modes of imaging employed in STMs. In (a), the tip–substrate separation is fixed and the variations in tunnel current I_t are related to variations in d. In (b), the tunnel current I_t is held constant by a feedback loop and the relative tip-substrate separation is varied to maintain a constant tunnel current.

the tip is moved at a fixed height above the substrate while variations in tunnel current due to height variations are recorded and (b) constant current imaging in which the tip position is continually adjusted by a feedback loop to produce a constant tunnel current. These two modes of imaging in STM are illustrated in Fig. 1.2.

The constant height mode illustrated in Fig. 1.2(a) is of limited use unless the sample surface is atomically flat since the tip position is not dynamically adjusted. In this mode of operation, the current variation is recorded as the tip is scanned across the substrate. This mode is most appropriate for substrates that are flat at the atomic length scale. Furthermore, since variations in tunnel current measured in the constant height mode depend exponentially with distance, observed variations in I_t cannot be directly interpreted as height profiles. The 3-dimensional imaging capability of STM is most easily understood by considering the constant current imaging mode shown in Fig. 1.2(b). An STM image results when the relative motion of the substrate is recorded while maintaining a constant tunnel current I_t as the tip is swept across a pre-selected area of the substrate. To achieve this, a high-gain current amplifier (typical gain is $\sim 10^8$ to $\sim 10^9$ V/A) is required.

Sharp tips are necessary to produce images with high lateral resolution. Common ways of producing STM tips from metal wires (such as W or Pt) with diameters of $\sim 1 \times 10^{-4}$ m rely on electrochemical etching or physical cutting, followed by thermal annealing and sharpening in ultra-high vacuum. The reliable formation of sharp tips may seem like a daunting venture, but ultimately every tip formed must end with one (or possibly a few) atoms which ever so slightly protrude from the apex, forming a small mini-tip at the tip's apex. The presence of such mini-tips, along with the strong exponential drop-off of current with distance, provides a sensible way to understand why the total tunnel current between tip and substrate is dominated by an atomically small protrusion from an otherwise large tip.

Complete theories of STM have shown that the tunnel current can be related to the quantum wavefunction overlap between electron states in the tip and electron states in the substrate. This implies that the images obtained from an STM contain not only surface topographic information, but also information about the variation of the local density of electronic states. This complication provides a caveat against the interpretation of relative tip–substrate separation into surface topographic features. STM images are notable for the amazing detail they reveal about the atomic periodicity and surface morphology of clean, electronically conducting substrates.

1.4   The Atomic Force Microscope

While STMs provide a quantitative map of surface topography with atomic resolution, they suffer from a fundamental limitation that the substrate studied must be sufficiently conducting to support a tunnel current, a limitation that was recognized early in the development of the STM. In order to overcome this difficulty, an AFM was first demonstrated in 1986 by Binnig, Quate and Gerber [binnig86]. The operation of an AFM relies on the surface forces acting on a sharp tip in close proximity to a surface, a topic that will be discussed in detail in the coming chapters. These surface forces are ubiquitous and exist between tips of any material and substrates of any material. From the very beginning, AFM promised to solve the problem of atomically-resolved images of insulating substrates.

For sufficiently small tip–substrate separations, these interaction forces can range from 10's of pN to 10's of μN, with typical values of a few tens of nN. An understanding of these interaction forces is central to understanding how an AFM functions. Most importantly, these forces are not predicated on the fact that either the tip or substrate be electrically conducting. Because of the long-range nature of the interaction forces, the vertical resolution of an AFM is typically less than a nanometer (comparable to an STM) while the lateral resolution is determined by the tip radius and sample roughness and is generally somewhat larger than that for STMs. In contrast to STM, which focusses on ultra-clean surfaces that are atomically flat, AFMs are used to study a wide variety of different substrates — both rough and smooth. AFM images typically span an area ranging from a $\sim$100 nm to $\sim$10's of μm.

In practice, the operation of an AFM relies on a sharp tip that is usually supported on the end of a microcantilever whose minute deflections can be carefully monitored. As shown in Fig. 1.3, when a microcantilever with spring constant k_c (units of N/m) positions a tip distance z from a substrate, the cantilever will deflect toward the substrate by an amount q due to attractive interaction forces that exist between the tip and substrate.

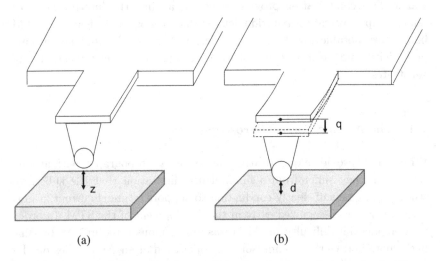

(a)   (b)

Fig. 1.3   A schematic illustrating the sequence of events when a tip on a microcantilever is brought into close proximity to a substrate. In (a), the apex of the tip is located a distance z above the substrate. Attractive surface interaction forces between the tip and substrate bend the tip toward the substrate until a deflection q of the cantilever brings the system into equilibrium. The final tip-substrate separation is indicated by the parameter d.

Table 1.2 A few representative AFM cantilevers commercially available with their characteristic dimensions, spring constants, and resonant frequencies.

material	length	width	thickness	k_c (N/m) (min, typical, max)	f_o (kHz) (min, typical, max)
Si	$125 \pm 5\,\mu$m	$35 \pm 3\,\mu$m	$4.0 \pm 0.5\,\mu$m	20, 40, 75	265, 325, 400
Si	$230 \pm 5\,\mu$m	$40 \pm 3\,\mu$m	$7.0 \pm 0.5\,\mu$m	25, 40, 60	150, 170, 190
Si	$90 \pm 5\,\mu$m	$35 \pm 3\,\mu$m	$2.0 \pm 0.3\,\mu$m	6.5, 14, 28	240, 315, 405
Si	$125 \pm 5\,\mu$m	$35 \pm 3\,\mu$m	$2.0 \pm 0.5\,\mu$m	1.8, 5.0, 12.5	110, 160, 220
Si	$90 \pm 5\,\mu$m	$35 \pm 3\,\mu$m	$1.0 \pm 0.3\,\mu$m	0.45, 1.75, 5.0	95, 155, 230
Si	$300 \pm 5\,\mu$m	$35 \pm 3\,\mu$m	$1.0 \pm 0.3\,\mu$m	0.01, 0.05, 0.1	9.5, 14, 19

For sufficiently small deflections, the cantilever motion can be well approximated in terms of Hook's law, which predicts an upward restoring force proportional to the cantilever deflection q given by

$$F_{restoring} = -k_c q. \tag{1.5}$$

In equilibrium, this restoring force must be equal and opposite to the interaction force that caused the deflection.

Table 1.2 provides typical dimensions and relevant properties of a few microcantilevers that are commercially available for use in AFM applications. Any uncertainty in cantilever thickness can cause considerable uncertainty in the resulting spring constants. Sharp tips, with effective radius R (typically R is between 5 and 30 nm), are routinely fabricated onto these cantilevers using lithographic techniques in common use by the microelectronic semiconductor industry. The wide-spread availability of suitable microcantilevers has enabled the routine measurement of interaction forces of order 1 nN between tip and substrate. Typically this means that cantilever deflections of order 1 nm or less can be detected.

To measure cantilever motion while scanning, a high-gain transducer of cantilever deflection plus a feedback mechanism is required. A variety of techniques — capacitance, optical interferometery, piezoelectric microcantilevers, and optical beam deflection — have been successfully implemented to accurately detect cantilever deflection. Each technique seems to have its own advantages. At this point in time, the technique most often implemented is an optical deflection scheme shown schematically in Fig. 1.4.

Using this approach, a laser is focused on the cantilever and the reflected light is directed onto a segmented photodiode. Fine positioning of the

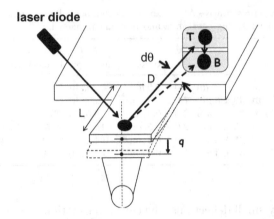

Fig. 1.4 A common method employed to measure the deflection of a cantilever is a beam bounce technique in which a diode laser beam is reflected from a microcantilever onto a segmented photodiode. By monitoring the voltage produced by the top (T) and bottom (B) segment of the photodiode, the relative motion of the reflected laser spot can be monitored and information about sub-nanometer deflection q of the cantilever can be inferred.

reflected spot on the photodiode allows for a null condition to be achieved. This occurs when the voltage from the top photodiode segment (T) equals the voltage from the bottom photodiode segment (B) in Fig. 1.4. A small cantilever deflection disrupts this null condition, giving rise to a voltage proportional to deflection. The origin of the high amplification for this particular system follows from simple geometrical considerations. For a cantilever displacement q, the reflected laser spot moves a distance Δs which is approximately given by

$$\Delta s \simeq q\frac{D}{L}, \tag{1.6}$$

where D is the distance of the cantilever from the photodiode and L is the cantilever's length. Typically, the ratio of D/L for an AFM microcantilever can easily be a factor of 100–500.

When discussing the nature of the interaction force between tip and substrate, it is often convenient to approximate the tip as a sphere with radius R as shown in Fig. 1.4. This sphere then interacts with the substrate via a number of possible forces that cause the cantilever to deflect.

The exact details of the relevant interaction forces, as well as their variations on d, depend to a large extent on the composition of the tip and substrate and will be further discussed in Chapters 2–4. For the ideal case

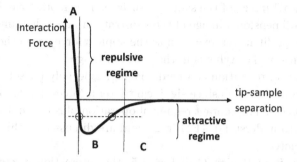

Fig. 1.5 A schematic illustrating how the interaction force between tip and substrate varies as a function of separation. Three regions (A, B, C) are indicated. Different modes of imaging are achieved when the tip is positioned in each region.

of a clean and electrically neutral tip positioned above a clean and electrically neutral substrate in ultra-high vacuum, the interaction forces might be well-approximated by a superposition of a short-range, hard-wall repulsion (effective when the tip–substrate separation is less than ∼0.3 nm) plus a longer range surface interaction due to the van der Waals (vdW) force acting between dipoles induced on the individual atoms comprising the tip and substrate. The variation of this interaction force on tip–sample separation is related to the detailed shape of the substrate and tip. If the system is operated under ambient air conditions, hydration forces due to adsorbed water or long-range electrostatic forces due to uncontrollable charging of the tip or substrate may well dominate.

Without a detailed knowledge of the system under study, it is difficult *a priori* to accurately specify a force vs. distance relationship. In general, such an interaction might be expected to follow the approximate shape shown in Fig. 1.5. This figure qualitatively illustrates (i) the attractive regime ($F < 0$) in which the interaction forces cause the microcantilever to bend toward the substrate and (ii) the repulsive regime ($F > 0$) that causes the microcantilever to bend away from the substrate when the tip comes into contact with it.

If the tip is in region A in Fig. 1.5, then imaging is performed in *contact mode*; the tip exerts a force directly on the sample as it is scanned across it. In contact mode imaging, the direct up and down motion of the cantilever is measured while scanning. A 3-dimensional image is built up as the sample is rastered beneath the tip. Usually, a feedback circuit raises and lowers the sample in such a way that the cantilever deflection remains constant. The amount the sample is raised or lowered at each (x, y) position forms the "topographical" image of the sample's surface. This motion can be used to produce a 3-dimensional image of the substrate in much the same way as a conventional profilometer, except now the applied force lies in the nN range and the radius of the stylus is in the 5–30 nm range.

This mode of operation can be damaging, especially for soft substrates and stiff microcantilevers since significant lateral forces develop during the scanning process. The contact regime of operation plus the underlying topics required to understand AFM in general are discussed in this volume of the lecture notes.

If the tip is in Region C of Fig. 1.5, the interaction forces are sufficiently weak so that very small deflections of the cantilever result. Since the substrate–tip separation is relatively large, imaging in this region is often referred to as the *non-contact mode*. Under these circumstances, indirect detection schemes are usually employed. As an example in non-contact mode imaging, the tip is often driven sinusoidally at a frequency near its mechanical resonance. Small position-dependent shifts in the resonance frequency then occur as the substrate is rastered beneath the tip. These frequency shifts provide a sensitive measure of tip–substrate interaction, thereby providing an input signal for a suitable feedback controller that adjusts the substrate position to maintain a constant frequency. The adjustments in the substrate position are then interpreted as a 3-dimensional topographic map of the surface structure. Because of the low forces applied to the sample, this mode is often preferred when studying soft substrates.

If the tip lies in Region B of Fig. 1.5, the slope of the interaction force becomes comparable to the restoring force of the microcantilever, implying that static tip displacements, although measurable, may not be reliable because of resulting instabilities. The instabilities arise because of the double-valued nature of the interaction force as illustrated schematically by the horizontal dotted line in Fig. 1.5, which shows that for the same value of the interaction force, there are two possible tip–sample separations. These instabilities are often referred to as jump-to-contact because the tip spontaneously snaps into contact with the substrate no matter how careful an approach procedure is followed. To scan in region B, the tip must be driven in a sinusoidal motion that is accurately monitored by the AFM control electronics. During the tip's motion, the tip will periodically come into contact with the substrate (Region A in Fig. 1.5), giving rise to what is known as *intermittent-contact* or *tapping mode* imaging. A careful analysis of the tip motion in this regime relies on solving the appropriate non-linear differential equations, a topic that will be covered in the 2nd volume of the lecture notes *Fundamentals of Atomic Force Microscopy, Part II: Dynamic AFM*. The added complication due to a non-linear tip motion has the advantage that property maps of sample stiffness, adhesion, etc. can be correlated with 3-dimensional topographic images.

The boundaries between the different regions in Fig. 1.5 are not necessarily well-defined. There is considerable discussion of these different regimes in the literature, so a precise distinction between them is challenging. For our purposes, this discussion might best be left to the opinion of experts. Suffice it to say, when the tip is oscillating, all Regions A, B and C are probed, and the AFM is often then referred to as a Dynamic Force Microscope (DFM). The imaging process is often referred to as dynamic mode imaging.

It is worth mentioning that a less complicated way of implementing intermittent contact mode imaging is to use an alternative approach often referred to as *jumping mode*. In this procedure, the cantilever does not undergo sinusoidal motion but instead follows a motion determined by software controlling the AFM. In practice, the software is programmed to position the tip at a distance far from the substrate, and then drive the substrate toward the tip under feedback control until the cantilever bending reaches a preset loading force. At this point the z-displacement of the substrate required to meet this preset loading condition is measured and the tip is withdrawn, moved to a nearby adjacent location above the substrate where the process is again repeated.

Upon completion of each 'jump', various features of the cantilever displacement as a function of the z-motion of the substrate are extracted from the data and plotted for further analysis. The advantage of this technique is that the force acting on the tip can be carefully monitored during the process by recording what is commonly called force vs. distance data. Furthermore, the lateral force imparted to the substrate while scanning is eliminated. The disadvantage is that a 'jump' image proceeds at a somewhat slower rate than when the cantilever is sinusoidally driven.

Figure 1.6 provides a schematic summary diagram of these different imaging modes.

Each of these modes of imaging requires different measurement techniques and control electronics. In addition, the mechanical properties of the cantilever must be selected for each technique to optimize success.

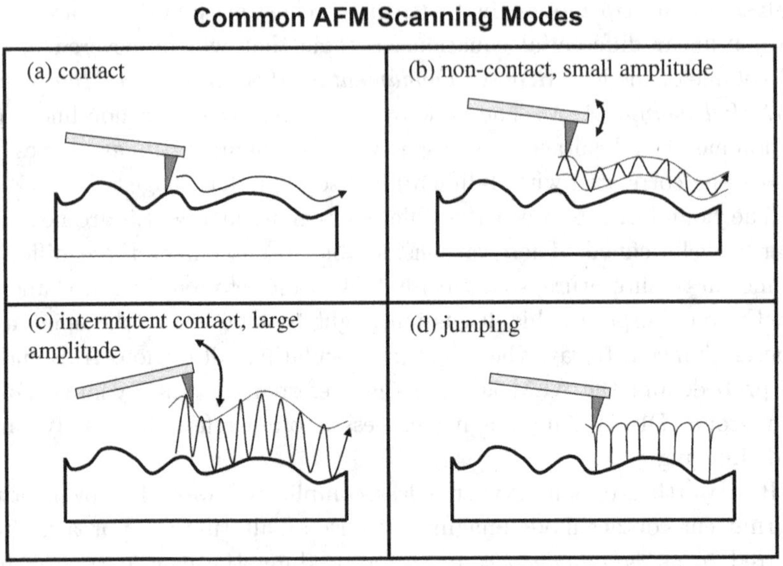

Common AFM Scanning Modes

Fig. 1.6 A schematic illustrating the different scanning modes commonly employed in AFM. In (a), the contact mode of imaging where the tip is in constant contact with the substrate. In (b), the non-contact mode of imaging, where the tip oscillates sinusoidally with small amplitude while maintaining a fixed distance between tip and substrate. In (c), the intermittent contact mode where the tip "taps" the substrate during the scanning process. The amplitude of the tip oscillation is now typically larger than in (b). Note that the frequency of tip oscillation in (b) and (c) is much greater than the scanning frequency of the microscope. In (d), the "jump" mode where the tip makes physical contact with the substrate, is then lifted and moved to another location before contact with the substrate is reestablished.

1.5 Current Trends in Atomic Force Microscopy

The basic techniques outlined above have been extended in a number of interesting ways, producing a large family of SPMs each specially tailored to detect the local variation in some quantity of interest. This extension of SPM is sometimes referred to as dual probe microscopy because the tip not only measures topography but also some other physical parameter of interest with high lateral resolution. A few examples include an electrostatic force microscope (EFM or scanning Kelvin probe), a magnetic force microscope (MFM), a photon scanning tunnelling microscope (PSTM), a scanning electrochemical microscope (SECM), a scanning near-field optical microscope (SNOM), a scanning capacitance microscope (SCM), scanning tunnelling spectroscopy (STS), and a frictional force microscope (FFM).

Current trends seek to exploit operation of AFM under liquid, to both image and probe soft biological material. Ever faster scanning requires ever smaller cantilevers that can oscillate more rapidly, pushing the upper frequency limits of detection electronics into the MHz regime. Spatial-dependent property maps are also collected in which not only the amplitude, but also the phases of the cantilever motion relative to the cantilever driving force, are measured and interpreted. Lastly, the cantilever is no longer treated as a simple oscillating thin beam, but as a dynamic vibrating object in which higher modes of oscillation are monitored to obtain spatial dependent property maps of substrates. Higher harmonic imaging has also become popular to reconstruct the nature of the interaction force vs. cantilever tip position.

There are many books and articles that have been written to summarize these advanced developments and a listing of useful references is included at the end of this chapter.

1.6 Chapter Summary

The rapid evolution of SPMs since their first demonstration in the early 1980s has truly been remarkable. Largely because they are versatile and relatively inexpensive, SPMs have ushered in a world-wide interest in nanotechnology. SPMs are notable because they provide high resolution (often atomic scale) metrology. Also, it is now clear that SPM tips can be used as tools capable of nanometer manipulation and fabrication. As an example, AFM nanolithography utilizes an AFM tip to locally modify a substrate in a very precise way.

SPM operation has been extended to scanning under liquid, allowing a window into the biological world. Advances in high-speed scanning are rapidly occurring, indicating that smaller cantilevers with higher resonant frequencies may lie in the near future. Linear parallel arrays of cantilevers have been fabricated to work in a massively parallel fashion and efforts to independently control individual cantilevers in the array have also been reported. Current indicators are that technology underlying these proximal probe microscopes will continue to improve, and the scanning probe class of instruments will continue to become ever more commonplace as a tool of choice to probe the properties of nanoscale objects.

When using an AFM, no matter which mode of imaging is employed, the motion of the substrate required to keep a relevant voltage signal constant at some pre-set value is used to form an image. In contact mode imaging, a voltage proportional to the static deflection of the cantilever is used. In non-contact mode imaging, either the amplitude or frequency of the cantilever oscillation is employed as a feedback signal.

The popularity of AFM as a core technique for surface metrology and characterization on a wide variety of different samples is now well documented. Using intermittent contact mode imaging, atomically resolved AFM images have been demonstrated that match the resolution of STM [giessibl03] [sugimoto07] [sugimoto07], but this usually requires an AFM operating under ultra-high vacuum conditions. Recent work has also demonstrated atomic resolution under water [fukuma05], [melcher13]. Under controlled conditions, the vertical resolution of AFM (typically better than 1 nm) rivals that of STM, while a force sensitivity of order 1 pN can be achieved under skilful operation.

1.7   Further Reading

Chapter One References:

[binnig82]   G. Binnig, H. Rohrer, Ch. Gerber, E. Weibel, "Tunnelling through a controllable vacuum gap", Appl. Phys. Lett., **40**, 178–80 (1982).

[binnig83a]   G. Binnig, H. Rohrer, C. Gerber, E. Weibel, "7 × 7 Reconstruction of Si(111) Resolved in Real Space", Phys. Rev. Lett. **50**, 120–23 (1983).

[binnig83b]   G. Binnig and H. Rohrer, "Scanning tunnelling micro-scopy", Surf. Sci. **126**, 236–244 (1983).

[binnig86]   G. Binnig, C.F. Quate, C. Gerber, "Atomic Force Microscope", Phys. Rev. Lett. **56**, 930–33 (1986).

[binnig87]   G. Binnig and H. Rohrer, "Scanning Tunnelling Microscopy — from Birth to Adolescence (Nobel Lecture)", Angewandte Chemie **26**, 606–614 (1987).

[fukuma05]   T. Fukuma, K. Kobayashi, K. Matsushige, H. Yamada, "True atomic resolution in liquid by frequency-modulation atomic force microscopy", Appl. Phys. Lett. **87**, 034101 (2005).

[giessibl03]   F.J. Giessibl, "Advances in atomic force microscopy", Rev. Mod. Phys. **75**, 949 (2003).

[gomez05]   Several sections in this chapter are adapted from J. Gomez-Herrero and R. Reifenberger "Scanning Probe Microscopy" in *Encyclopedia of Condensed Matter Physics*, eds. F. Bassani, J. Leidl, P. Wyder, Elsevier Science Ltd. pgs. 172–82 (2005).

[haguenau03]   F. Haguenau, P.W. Hawkes, J.L. Hutchison, B. Satiat-Jeunemaître, G.T. Simon and D.B. Williams "Key Events in the

History of Electron Microscopy", Microsc. Microanal. **9**, 96–138 (2003).

[melcher13]   J. Melcher, D. Martínez-Martín, M. Jaafar, J. Gómez-Herrero and A. Raman, "High-resolution dynamic atomic force microscopy in liquids with different feedback architectures", Beilstein J. Nanotechnol. 2013, 4, 153–163.

[muller56]   E. Müller and K. Bahadur, "Field Ionization of Gases at a Metal Surface and the Resolution of the Field Ion Microscope". Phys. Rev. **102**, 624 (1956).

[muller65]   E.W. Müller, "Field Ion Microscopy", Science **149** (3684), 591–601 (1965).

[shmalz29]   G. Shmalz Über Glätte und Ebenheit als physikalisches und physiologisches Problem, Vereines Deut. Ingen. (VDI), **73**, 1461–67 (1929).

[sugimoto07]   Y. Sugimoto, P. Pou, M. Abel, P. Jelinek, R. Pérez, S. Morita, Ó. Custance, "Chemical identification of individual surface atoms by atomic force microscopy", Nature **446**, 64 (2007).

[tipler12]   P.A. Tipler and R.A. Llewellyn, *Modern Physics, 6th edition*, (W.H. Freeman and Co., New York), 2012.

[young71]   R. Young, J. Ward and F. Scire, "Observation of metal-vacuum-metal tunnelling, field emission, and the transition regime", Phys. Rev. Lett. **27**, 922 (1971).

[young72]   R. Young, J. Ward and F. Scire, "The topographiner: An instrument for measuring surface microtopography", Rev. Sci. Instrum., **43**, 999 (1972).

Further reading:

[baro12]   *Atomic Force Microscopy in Liquid: Biological Applications*, Eds. A.M. Baro and R. Reifenberger, (Wiley-VCH, Berlin), 2012.

[bhushan]   B. Bhushan and H. Fuchs, *Applied Scanning Probe Methods Vols. I–XIII* (Springer, Berlin).

[binnig99]   G. Binnig and H. Rohrer, "In touch with atoms", Rev. Mod. Phys. **71**, S324–S330 (1999).

[cappella99]   B. Cappella and G. Dietler, "Force distance curves by atomic force microscopy", Surf. Sci. Rep. **34**, 1–104 (1999).

[chen93]   C. Julian Chen, *Introduction to Scanning Tunnelling Microscopy*, Oxford series in Optical and Imaging Sciences, Oxford University Press, New York (1993).

[garcia02]   R. Garcia and R. Perez, "Dynamic atomic force microscopy methods", Surf. Sci. Rep. **47**, 197–301 (2002).

[guntherodt94]   *Scanning Tunneling Microscopy I: General Principles & Applications to Clean & Adsorbate-Covered Surfaces*, Eds. H.J. Guntherodt and R. Wiesendanger, Springer-Verlag. New York (1994).

[guntherodt97]   *Scanning Tunneling Microscopy II: Theory of STM & Related Scanning Probe Methods*, Eds. H.J. Guntherodt and R. Wiesendanger, Springer-Verlag, New York (1997).

[marti93]   *STM and SFM in Biology*, Eds. O. Marti and M. Amrein, Academic Press, New York (1993).

[meyer04]   E. Meyer, H.J. Hug, R. Bennewitz, *Scanning Probe Microscopy*, Springer, Berlin (2004).

[sarid91]   D. Sarid, *Scanning Force Microscopy with applications to electric, magnetic, and atomic forces*, Oxford Series in Optical and Imaging Sciences, Oxford University Press, New York (1991).

Chapter 2

The Force between Molecules

Before discussing the operation of an AFM, we need to understand more about the nature of forces at the molecular length scale since these forces ultimately govern tip–substrate interactions. As we shall see, an understanding of these inter-molecular forces requires cross-disciplinary knowledge derived from both chemistry and physics.

What outlook should you adopt when reading this chapter? In AFM, more often than not, we can't work out the exact details of a particular tip–substrate interaction — we simply do not have enough information about all constituents to make such a calculation possible. If this is the case, then why bother reading the discussion that follows? Usually it is possible to make educated guesses about an unknown system from experience gained by analyzing simpler, analytical models. The way forward is to develop an ability to make quantitative predications, which are then tested to check the consistency and validity of the original guesses. The insights gained from studying the model systems discussed in this chapter should sharpen your initial educated guesses.

To use an AFM intelligently, you should be aware of simple models to estimate inter-molecular forces and you should understand which atomic and molecular properties control these interactions. These topics are typically of most interest to students in chemistry and physics. If you are already familiar with electric fields and electrostatic potential energy, with molecular dipole moments and how they influence physical properties, and the calculation of dipole moments from molecular structures, you can skip over the discussion found in this chapter.

2.1   Evidence for Inter-Molecular Forces

An AFM senses the minute forces between a sharp tip and a nearby substrate. Ultimately, the net force between the tip and substrate is the sum of forces between individual atoms/molecules in the tip and atoms/molecules in the substrate. Since molecules are formed from atoms, we will use "atoms" in the following discussion rather than the awkward but more precise "atoms/molecules". Quantifying the forces between individual atoms is required for understanding the tip–substrate interaction force. Can we build some intuition about these forces that are extremely small (10^{-12} to 10^{-9} N) by everyday standards? What are the physical manifestations of these forces that can be measured in the laboratory?

Perhaps the first clear evidence for inter-molecular forces was the observation that all gasses showed a systematic deviation from ideal gas law behavior as the pressure increased. To account for these deviations, Johannes D. van der Waals proposed in 1873 a revised equation of state to relate the pressure P (Pa), the volume V (m^3) and the temperature T (Kelvin) for any gas [vdwaals04]

$$\left(P + \frac{a}{(\frac{V}{n})^2}\right)\left(\frac{V}{n} - b\right) = R_A T, \qquad (2.1)$$

where a (units of $N \cdot m/mole^2$), b (units of $m^3/mole$) are fitting constants that depend on the composition of the gas, n is the number of moles and R_A ($8.3145\, J\, mole^{-1}\, K^{-1}$) is the universal gas constant which is equal to $N_A k_B$, where N_A is Avagadro's number ($6.02 \times 10^{23}/mole$) and k_B is Boltzmann's constant with the value of $1.38 \times 10^{-23}\, J/K$.

The sum of the volumes of one mole of individual gas molecules is measured by the parameter b. The parameter a is a pressure correction term that accounts for attractive intermolecular interactions. In effect, if the gas molecules attract each other, then the collisions with the walls of a containing vessel are reduced, thus decreasing the pressure required to achieve a specified volume.

A second example for the existence of inter-atomic forces comes from comparing the boiling points of liquids. For a liquid to boil, sufficient heat must be added to overcome the inter-atomic (or inter-molecular) forces that are present. The fact that one pure liquid boils at a higher temperature than another is an indicator that different atoms are attracted by each other with different inter-molecular forces.

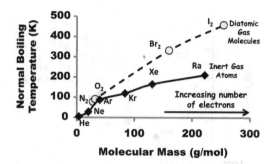

Fig. 2.1 Boiling points of monoatomic (inert gas) atoms and diatomic gas molecules showing a systematic increasing trend as the molecular mass increases.

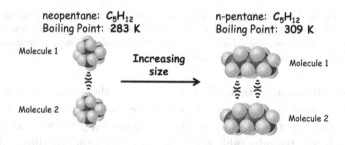

Fig. 2.2 The structural isomers of pentane have two different molecular shapes and two different boiling temperatures.

Figure 2.1 plots the boiling temperature under normal pressure for monoatomic gas atoms and diatomic molecules and shows that the boiling temperature tends to increase with the number of electrons comprising the atoms or molecules. This trend suggests that inter-molecular interactions generally increase with electron number.

A third example comes from considering structural isomers, molecules having the same chemical formula (the same atoms) that bind together in different ways to form different structures. For example, compare the boiling point of neopentane (2,2-dimethylpropane; C_5H_{12}) — a volatile, chain alkane with five carbon atoms — to its structural isomer n-pentane (also C_5H_{12}). As shown in Fig. 2.2, there is a 26 K change in the boiling temperature between pure liquids of these two isomers that must be attributed to differences in their shape. N-pentane with a linear molecular shape exhibits a stronger inter-molecular interaction than the more compact neopentane molecules.

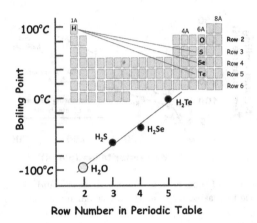

Fig. 2.3   The measured boiling point of H_2Te, H_2Se, H_2S. If H_2O follows the trend, the prediction for the boiling point of H_2O would be near $-100°C$. The actual boiling point of H_2O at $+100°C$ provides strong evidence of an anomalously strong inter-molecular interaction between water molecules [kotz06].

A fourth example that offers insight into the nature of the inter-molecular forces comes from comparing the boiling points of inorganic liquids formed by the chemical reaction of hydrogen with the chalcogenide elements in the periodic table (O, S, Se, Te). The resulting dihydrides are "bent-like" molecules of the type H-x-H where x represents O, S, Se, and Te. These molecules condense into liquids with a boiling point that systematically varies with row position in the periodic table as shown in Fig. 2.3. The bond lengths of the dihydride molecules also tend to increase down a chemical group (increasing row number). An interesting trend is observed in this graph that predicts the boiling point of water to lie near $-100°C$. The actual boiling temperature of water at $+100°C$ indicates an anomalously strong interaction between water molecules, which is attributed to hydrogen bonding.

Similar trends are observed for simple molecules containing N and F as indicated in Table 2.1 below, providing evidence that anomalously high inter-molecular interactions occur whenever O, N, or F atoms are incorporated into molecules.

These simple considerations provide experimental evidence that shed light on inter-molecular interactions. Taken together, one might expect "larger" molecules to have higher boiling points than smaller molecules of a similar type, indicating that attractive inter-molecular forces generally increase in some unspecified way with (i) molecular mass, (ii) the number of

Table 2.1   A compilation of boiling points for a series of various liquids. The presence of O, F, or N produces an anomalously high boiling temperature when compared to other values in the series [kotz06].

Chalcogens	Boiling point	Halogens	Boiling point	Pnictogens	Boiling point
H_2Te	$-2°C$	HI	$-35°C$	H_3Sb	$-17°C$
H_2Se	$-41°C$	HBr	$-67°C$	H_3AS	$-63°C$
H_2S	$-60°C$	HCl	$-85°C$	PH_3	$-88°C$
H_2O	$100°C$	HF	$20°C$	NH_3	$33°C$

electrons in a molecule, or (iii) the number of atoms comprising a molecule. The strength of the interaction also tends to increase drastically if O, N, F atoms happen to be involved in the chemical makeup of the molecular structure.

To better understand these trends and gain further insights at the microscopic level, we need to briefly review some standard results that are typically covered in an introductory university-level physics class. An understanding of these topics is required to better appreciate the intra-atomic forces that form molecules. These forces also provide a basis for understanding the origin of the inter-atomic forces discussed above.

2.2   A Review of Relevant Electrostatics

2.2.1   *Coulomb's law for point charges*

The most straightforward manifestation of a classical electrostatic force is the case of two point charges, a situation which is taught in every introductory first-year physics course under the title of Coulomb's Law. For two point charges q_1 and q_2 separated by a distance z in free space, the magnitude of the electrostatic force of interaction can be written as

$$|\vec{F}| = \frac{1}{4\pi\varepsilon_o} \frac{|q_1||q_2|}{z^2} \quad \text{(two point charges in vacuum)}, \quad (2.2)$$

where $\varepsilon_o = 8.85 \times 10^{-12} \, C^2/(N \cdot m)$ is the electrical permittivity of vacuum, the charges q_1 and q_2 are measured in coulombs (C), and the charge separation distance z is measured in meters (m). If the charges are embedded in a uniform dielectric, say a liquid having a dielectric constant κ (with dielectric permittivity $\varepsilon = \kappa\varepsilon_o$), the electrostatic force in Eq. (2.1) would

be written as

$$|\vec{F}| = \frac{1}{4\pi\kappa\varepsilon_o}\frac{|q_1||q_2|}{z^2}. \quad \text{(two point charges in dielectric)} \quad (2.3)$$

Equations (2.2) and (2.3) are macroscopic formulas in the sense that they apply only to charges embedded in materials which are homogenous in an average way. As written, they give the magnitude of the electrostatic force. Since force is a vector, it has a direction. For the case of two point charges, the direction is easy enough to specify by the simple rule: like charges repel, unlike charges attract.

It is clear that the factor of κ reduces the electrostatic force, providing an explanation for why liquids with high dielectric constants are often good solvents. Nearby ions of opposite polarity dissolved in the liquid feel a reduced force of attraction that hinders them from recombination and eventual agglomeration. When point charges are embedded in solid dielectrics, Eq. (2.3) should be used with caution [feynman64].

Rarely do we find only two point charges located in empty space. Typically, extended objects acquire charge distributions that can be described by linear charge densities λ (C/m), surface charge densities σ (C/m^2), or volume charge densities ρ (C/m^3). In this case, forces between objects require an integration taking into account the 3-dimensional distribution of the charge.

Example 2.1: What is the force exerted on a point charge Q placed at a distance z above the center of a thin flat plate of dimension $a \times b$ (see Figure below)? The plate has a uniform charge distribution σ.

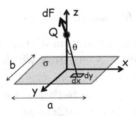

The problem requires an integration of the force dF between the charge q contained in an area $dA = dxdy$ and the charge Q. The separation between dA and Q is

$$\sqrt{x^2 + y^2 + z^2}.$$

The element of force dF has two components, one parallel to z, the other parallel to the plane. By symmetry, the force parallel to the plane

(Continued)

Example 2.1: (*Continued*)

will cancel. The element of force along z, dF_z, is equal to $|dF|\cos\theta$. The total force along z is given by integration. The charge in an area element $dx\,dy$ is given by

$$q = \sigma\,dx\,dy$$

$$|dF| = \frac{1}{4\pi\varepsilon_o}\frac{Qq}{(r)^2} = \frac{1}{4\pi\varepsilon_o}\frac{Q\sigma\,dx\,dy}{(x^2+y^2+z^2)}$$

$$dF_z = |dF|\cos\theta = |dF|\frac{z}{(x^2+y^2+z^2)^{\frac{1}{2}}}$$

$$F_z(z) = \frac{\sigma Q}{4\pi\varepsilon_o}\int_{-\frac{a}{2}}^{+\frac{a}{2}}dx\int_{-\frac{b}{2}}^{+\frac{b}{2}}dy\frac{z}{(x^2+y^2+z^2)^{\frac{3}{2}}}$$

$$= \frac{\sigma Qz}{4\pi\varepsilon_o}\int_{-\frac{a}{2}}^{+\frac{a}{2}}dx\int_{-\frac{b}{2}}^{+\frac{b}{2}}\frac{dy}{(x^2+y^2+z^2)^{\frac{3}{2}}}$$

$$= \frac{\sigma Q}{\pi\varepsilon_o}\left[\tan^{-1}\left(\frac{ab}{2z\sqrt{4z^2+b^2+a^2}}\right)\right]$$

The final answer is complicated and to better understand the result it is useful to make a log-log plot of $F_z(z)$ as shown below. Two situations are plotted: (i) a small uniformly charged square plate with dimension $a = b = 2\,\mu\text{m}$ and (ii) a rectangular plate with $a = 4\,\mu\text{m}$, $b = 200\,\mu\text{m}$. The total charge on the plate in both cases is σab.

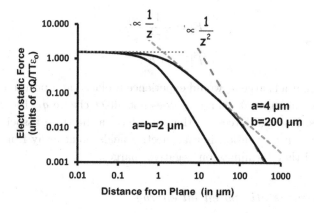

(*Continued*)

Example 2.1:   (*Continued*)

As z increases, the force varies with z in such a way that provides insight into the interaction.

When z approaches 0, the argument of the arc tangent grows without bound and $\tan^{-1} \to \pi/2$. The electrostatic force is constant and equal to that expected for a point charge above a flat plane having a constant surface charge density σ, namely $F = Q\sigma/2\varepsilon_o$. As z increases, the force decreases from this constant value.

For the case of the large rectangular plate ($a = 4\,\mu$m, $b = 200\,\mu$m), the net force exhibits two well-defined regimes. The two dashed lines indicate forces that vary as $1/z$ and $1/z^2$. When the force varies as $1/z$, the situation is approximately equivalent to the point charge Q interacting with a line charge (producing a $1/z$ variation). When $z \gg a$, $z \gg b$, the rectangular plate resembles a point charge that produces a force that varies as $1/z^2$.

This example is chosen to emphasize that the variation of force with separation is often a useful way to understand and classify the type of interaction between two objects.

Coulomb's Law is often discussed in terms of an electric field that develops around a particular distribution of charge. In the simplest case for two point charges separated by a distance z in vacuum, it is very convenient to define the magnitude of an electric field $\vec{E}$ generated by charge q_1 at a distance z from q_1 as

$$|\vec{E}| = \frac{1}{4\pi\varepsilon_o} \frac{|q_1|}{z^2}, \qquad (2.4)$$

$$|\vec{F}| = q_2|\vec{E}|. \qquad (2.5)$$

The force that a charge q_2 would experience if placed at a position z is then given by assuming that charge q_2 does not affect charge q_1. Electric fields are vectors, just like forces, so they also have a direction associated with them. The sign convention is that electric fields point away from positive charges and they terminate on negative charges.

2.2.2   *Electrostatic potential energy*

For two positive charges interacting with each other, an external force is required to overcome the electrostatic repulsive Coulomb force so that a

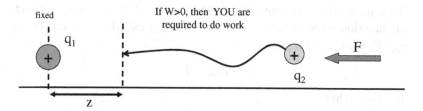

Fig. 2.4   Moving a charge q_2 in the presence of a fixed stationary charge q_1 requires an external force.

charge q_2 can be moved to a new location. The situation is indicated schematically in Fig. 2.4. In analogy with a mechanical problem requiring work to lift a mass m in a gravitational field, it is possible to define the electrostatic work required to place a charge q_2 at a distance z from the stationary charge q_1.

If the electrostatic force is conservative, then it is possible to define the work W (in Joules) to move an object from a reference point at infinity to some point as the line integral of the force component parallel to the displacement vector $d\vec{\ell}$ according to

$$W \equiv \int_{\infty}^{z} \vec{F} \cdot d\vec{\ell}. \qquad (2.6)$$

The work defined in this way is a signed quantity and the sign can be used to distinguish between work performed by an external agent ($W > 0$; work done on the system) and work performed by you ($W < 0$) in order to locate the charge at its final position. It is useful to briefly digress and consider the work performed in bringing a system to a final charge configuration (Fig. 2.4) within the context of the First Law of Thermodynamics. The First Law states

$$dE_{\text{int}} = TdS + dW, \qquad (2.7)$$

where dE_{int} is the change in the internal energy of the system, T is the absolute temperature of the system, dS is the change in entropy, TdS is the heat added to the system during the process under consideration, and dW is the work performed to assemble the system of charges.

As seen from Eq. (2.7), the work performed can be identified with a change in internal energy **only if** the work is performed adiabatically ($dS = 0$). Since a change in temperature accompanies any adiabatic process, this discussion is not particularly useful since one might expect an isothermal process ($dT = 0$) to characterize electrostatic charging.

To remedy this situation, it is customary to define a new thermodynamic function relevant for electrostatic charging called the Helmholtz free
energy F, which is defined as

$$F \equiv E_{\text{int}} - TS. \tag{2.8}$$

It follows that

$$dF = dE_{\text{int}} - TdS - SdT$$

$$= dW - SdT. \tag{2.9}$$

From Eq. (2.9), it is now easy to infer that the electrostatic work dW can
be equated with a change in the Helmholtz free energy dF if the charging is
done isothermally ($dT = 0$). From chemical thermodynamics, the standard
interpretation for a change in Helmholtz free energy is the maximum work
that can be extracted from a system at some later time. In the case of
electrostatic charging, we say the energy is stored in the electric field rather
than in the chemical bonds.

A useful quantity called the electrostatic potential energy $U(z)$ can be
defined by

$$U(z) \equiv -W. \tag{2.10}$$

The clear implication of Eq. (2.10) is that any work put into assembling
a system of charges can be stored as electrostatic potential energy $U(z)$,
which in principle can be recovered at a later time.

Calculating $U(z)$ for relevant charge distributions will become a central
focus of Chapters 3 and 4. The utility of $U(z)$ is that it directly measures the
energy required to assemble a specified charge configuration with respect
to a reference specified by the lower limit of the integral that appears in
Eq. (2.6). This reference is somewhat arbitrary but for the case of electrostatics, it usually corresponds to the case when all charges are infinitely far
apart.

Once $U(z)$ is known, the force acting on a charge located at a position
z can be calculated according to

$$F(z) = -\frac{\partial U}{\partial z}. \tag{2.11}$$

Equation (2.11) will be used extensively in the following chapters. Even
though this book is concerned primarily with forces at the atomic scale,
we will spend most of our time calculating energies and taking derivatives.
Even though we continually speak about intermolecular forces, it is usually

more convenient to calculate the interaction between molecules in terms of $U(z)$ and then take a derivative to find the force.

The results presented above for 1-dimension can be generalized to 3-dimension using vector calculus. Defining a point infinitely far away as a reference potential energy equal to 0, the electrostatic potential energy required to position a charge q at some point P can be defined as

$$U_P(\vec{r}) \equiv - \int_\infty^P q\vec{E}(\vec{r}) \cdot d\vec{\ell}. \qquad (2.12)$$

The force on the charge q at the point P is given by

$$\vec{F} = -\vec{\nabla}U(\vec{r} = P)$$

$$= - \left.\frac{\partial U(x,y,z)}{\partial x}\right|_P \hat{x} - \left.\frac{\partial U(x,y,z)}{\partial y}\right|_P \hat{y} - \left.\frac{\partial U(x,y,z)}{\partial z}\right|_P \hat{z}, \qquad (2.13)$$

where $\hat{x}, \hat{y}$, and $\hat{z}$ are understood to be unit vectors pointing in the x, y, and z directions. The utility of first calculating $U(r)$ and then taking the gradient to obtain the vector force $\vec{F}$ is an important procedural detail that should not be underestimated.

For two point charges q_1, q_2 separated by a distance z in vacuum, $U(z)$ has the familiar analytical form given by

$$U(z) = \frac{1}{4\pi\varepsilon_o} \frac{q_1 q_2}{z}. \qquad (2.14)$$

$U(z)$ is clearly a signed quantity that depends on the polarity of the two charges q_1 and q_2.

Example 2.2: Plot the electrostatic potential energy of two point charges separated by a distance z. Assume a point charge $+Q$ is fixed at the origin. Consider two cases: (i) when the moveable point charge q is positive, and (ii) when the moveable point charge q is negative.

A plot of the two cases is given schematically in the figure below, which shows how $U(z)$ increases with z for the case when q is positive and how $U(z)$ decreases with z for the case when q is negative. The starting and ending locations of the charge q are also indicated.

(Continued)

Example 2.2:   (*Continued*)

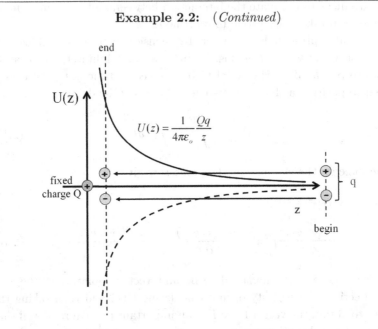

In general, if $U(z)$ increases as a result of the motion, an external force is required to push the charge q to its final location. The final configuration is then said to be repulsive. On the other hand, if $U(z)$ decreases as q approaches Q, the final configuration is then said to be attractive.

It is important to understand this sign convention since it will be used extensively throughout our discussions.

2.3   The Forces that Hold Molecules and Solids Together

Understanding the forces that hold molecules together requires a general appreciation for the origin of bonding in molecules and solids. This topic is well-established with a rich history and we make no attempt to systematically discuss it here [pauling32] [mulliken34]. For our purposes, the main results can be simplistically summarized in Fig. 2.5, which illustrates that chemical bonding varies smoothly along a continuum, rather than being sharply divided into well-defined categories such as covalent and ionic. Chemical bonding ranges from equal electron sharing between identical

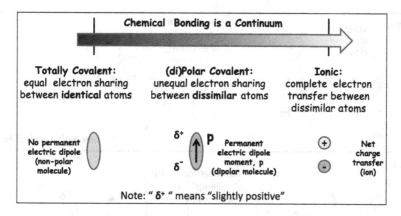

Fig. 2.5 The chemical bond that holds atoms together forms a continuum that spans the range shown above. On one extreme is an equal sharing of electrons between identical atoms to form covalent bonds. At the other extreme is a complete electron transfer between dissimilar atoms to form ionic bonds.

atoms (H_2 molecule) to the total electron transfer between atoms to form ionic compounds like NaCl.

At the atomic level, whenever electrons are confined to small dimensions they can no longer be treated as a point charge. Instead they must be viewed as a delocalized wave characterized by a quantum mechanical wavefunction. The quantum wavefunction derived from Schrödinger's equation provides the framework for describing the shape of the electronic charge cloud. Rather than reviewing the assortment of different electron wave functions, we assume the shapes of the electron orbitals are familiar from introductory chemistry courses.

To describe the continuous nature of the chemical bond it becomes necessary to describe the charge distribution within a molecule. A schematic is shown in Fig. 2.6 for two cases. The redistribution of electronic charge when the two atoms are far apart is contrasted to the case when the atoms are close together at their equilibrium spacing. The electron wavefunctions are schematically indicated as uniform spherical charge clouds in this figure.

When the atoms involved are identical, the final electron charge distribution must reflect this symmetry and hence the charge distribution must lie along the bisector between the two nuclei and be symmetrically distributed about it. When the two atoms involved are dissimilar, one atom will inevitably attract electrons more than the other, resulting in a final electron charge distribution that is asymmetric along the bisector between the two nuclei.

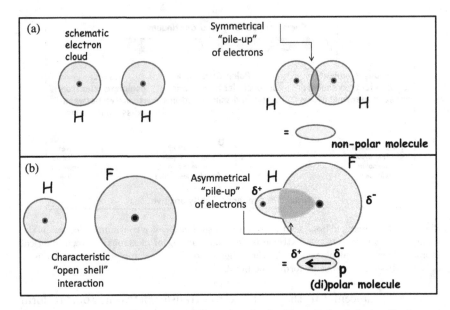

Fig. 2.6   A schematic illustration of distortions in the electron cloud during the formation of non-polar and dipolar covalent molecules. Any dipole moment p that develops provides a useful way to characterize the interaction between molecules.

It is useful to identify atoms that strongly attract electrons because molecules comprised of those atoms are likely to be highly dipolar. The well-known electronegativity series for the elements provides this ranking as shown in Fig. 2.7. The chart is useful for qualitatively assessing the relative electronegativity of various elements one with respect to another.

Example 2.3: Hydrocarbons are organic compounds consisting entirely of hydrogen and carbon. Would you expect hydrocarbon molecules to have a large dipole moment?

By inspection of Fig. 2.7, the positions of H and C in the electronegativity chart lie roughly at the same vertical location. Therefore, neither atom is more electronegative than the other. As a result, you might expect hydrocarbon molecules to have relatively small dipole moments.

Molecules with small dipole moments will not strongly interact with one another. As a consequence, when hydrocarbon molecules condense into the liquid state, you should expect the liquid to have a relatively high vapor pressure which leads to rapid evaporation.

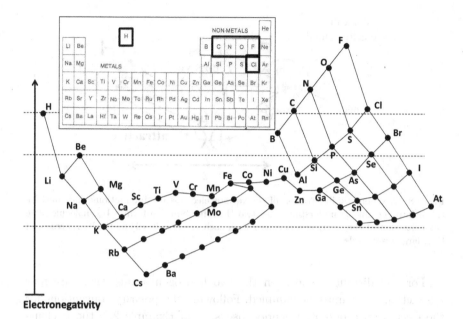

Electronegativity

Fig. 2.7   A chart of the relative electronegativity of elements in the periodic table. Elements near the top of the chart are most electronegative and have the highest affinity for attracting electrons. The most electronegative elements are F, O, N, and H, C and Cl.

Source: http://users.rcn.com/jkimball.ma.ultranet/BiologyPages/E/Electronegativity. html Last accessed December 24, 2013.

2.4   Electrostatic Forces Lead to Stable Molecules

The electrostatic forces between atoms can result in the formation of stable molecules. Making precise predictions regarding the stability of a molecule requires detailed quantum mechanical calculations, which are necessary to properly describe the charge distribution accurately. It is useful to think through a simple example in order to understand the important concepts.

Consider the simplest molecule H_2 in which two H atoms are separated by an adjustable distance z. The H atom is represented by a positive point charge nucleus and a negatively charged quantum mechanical electron cloud, represented schematically by a spherical charge distribution as shown in Fig. 2.8. The shape of the electron charge distribution must be self-consistently calculated for every separation distance z. For the case of the H_2 molecule, the charges redistribute as shown schematically in Fig. 2.6.

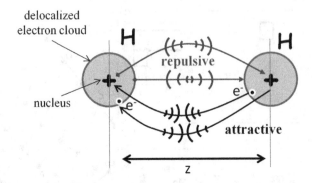

Fig. 2.8   A schematic illustration of the various electrostatic interactions that must be taken into account to understand how two H atoms bond to form a H_2 molecule. The delocalization of the electron associated with each H atom is schematically represented by a large grey circle.

For any distance z between the two hydrogen atoms, there are four interactions that must be summed. Following the polarity conventions for the electrostatic potential energy discussed in Example 2.2, the nucleus–nucleus and the electron–electron interactions between H atoms are repulsive (like charges repel). On the other hand, the nucleus–electron interaction between the two H atoms is attractive (unlike charges attract). The net electrostatic potential energy will depend on the relative strength of these four contributions.

An important result is that as z is decreased, the sum of the four contributions changes from negative (attractive) to positive (repulsive). Initially, for large separations, the net electrostatic potential energy is near zero. As z decreases and the two H atoms are brought closer together, the electrostatic potential energy decreases since energetically it becomes favorable for the electron charge cloud to be localized at the midpoint between the two nuclei, essentially shielding the repulsive nuclear-nuclear interaction. As z continues to decrease even more, the electrostatic interaction eventually becomes positive due to the dominant repulsive interactions.

An empirical model that describes this broad class of interactions between two electrically neutral atoms is known as the 12-6 Lennard–Jones interaction potential, which is often written as

$$U(z) = 4\varepsilon \left\{ \left(\frac{\sigma}{z}\right)^{12} - \left(\frac{\sigma}{z}\right)^{6} \right\}. \tag{2.15}$$

Equation (2.15) contains two adjustable parameters, which can be used to model the interaction: ε controls the depth of the minimum in $U(z)$ and σ controls the separation at which $U(z) = 0$. Although the 12-6 Lennard–Jones potential is widely used because of its numerical simplicity, it is known to have serious defects, especially at both small and large values of z. It is perhaps of most use to model general trends rather than specific properties of a system.

A plot of the Lennard–Jones potential as a function of the nuclear separation distance z is given in Fig. 2.9 with parameters commonly used to model Ar–Ar interactions. The minimum in $U(z)$ describes the most energetically favorable location for the two atoms and coincides with the condition of no net force.

It is important to realize that different regions of the Lennard–Jones potential provide information about different processes. As an example, Region I in Fig. 2.9 is most important in high pressure situations, the shape of Region II is most important when discussing equilibrium/vibrational problems, and Region III is most appropriate for describing high temperature, thermal expansion situations. The point to remember is that different regions of the interaction potential are responsible for different physical effects. This principle will be mirrored time and again when we consider AFM tip–substrate interactions in later chapters.

For the case of two Ar atoms at room temperature, thermal energies $k_B T$ (~25 meV) exceed the depth of the potential energy well plotted in Fig. 2.9, which is about 10 meV deep, indicating that two Ar atoms will

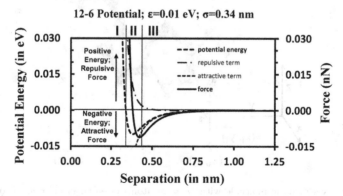

Fig. 2.9 The electrostatic potential energy $U(z)$ (Eq. (2.15), dotted line) and the resulting force $F(z) = -\partial U/\partial z$ (solid line) for two Argon atoms as a function of their separation distance z. Three regimes I, II, and III are schematically indicated.

not bind to each other at these temperatures. One might expect significant Ar–Ar interactions to occur when $k_B T \approx 10\,\text{meV}$, at a temperature near 115 K. Indeed, as the temperature is lowered, the thermal energy available reduces and Ar eventually condenses into a liquid around 86 K.

2.5   Molecular Dipole Moments

A calculation of the dipole moment of a molecule requires knowledge of the position of each atom along with the electronic charge distribution within the molecule. While the positions of the atomic nuclei are localized, the electron charge is continuously distributed throughout the molecule and must be obtained from a detailed quantum mechanical calculation. A convenient approximation is to associate a partial charge (either positive or negative) with the position of each nucleus as illustrated in Fig. 2.10. The localized partial charge is fictitious, since calculations provide the wave functions which can be used to calculate the charge density and the electrostatic potential, both of which are continuous quantities. The partial charge is often empirically determined by adjusting the charge on each atom to most closely match the electrostatic potential, using a least squares fit criterion.

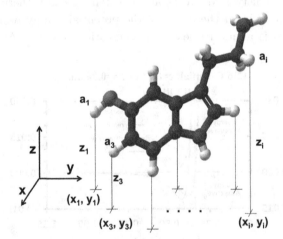

Fig. 2.10   A schematic molecule with $N = 24$ atoms oriented in an x, y, z coordinate system. It is assumed that the coordinates (x_i, y_i, z_i) and the residual charge (a_i) on each atom is known.

The partial charge δq_i on the ith atom can be specified by a quantity a_i such that

$$\delta q_i = a_i |e^-| \quad -1 \leq a_i \leq +1, \tag{2.16}$$

where the limits on a_i are imposed by the constraint that the charge transfer is assumed to be less than one electron. The dipole moment along x, y and z can then be calculated using

$$p_x = \sum_i x_i \delta q_i \quad p_y = \sum_i y_i \delta q_i \quad p_z = \sum_i z_i \delta q_i. \tag{2.17}$$

The magnitude of the dipole moment is then given by

$$|\vec{p}| = \sqrt{p_x^2 + p_y^2 + p_z^2}. \tag{2.18}$$

Does the above calculation for p depend on the definition of the coordinate origin? It is easy to show that if the original coordinate origin is shifted to a new location (x_o, y_o, z_o), the calculation of the dipole moment remains invariant as long as the condition

$$\sum_{i=1}^{N} a_i = 0. \tag{2.19}$$

is satisfied. This is equivalent to specifying an electrically neutral molecule with no net positive or negative charge. Historically the standard unit of a dipole moment was defined by two charges of opposite polarity but with equal magnitude of $\delta q = 1 \times 10^{-10}$ statCoulomb (also called an e.s.u. of charge — the electrostatic unit, in older literature), separated by a distance of 1 Angstrom $= 1.0 \times 10^{-8}$ cm. This leads to the unit of a Debye for atomic or molecular dipole moments:

$$1\,\text{Debye} \equiv \delta q \times d = (1 \times 10^{-10} \text{statC})(1 \times 10^{-8}\,\text{cm})$$

$$\times\,(3.33564 \times 10^{-10}\,\text{C/statC})$$

$$= 3.33 \times 10^{-28}\,\text{C} \cdot \text{cm} = 3.33 \times 10^{-30}\,\text{C} \cdot \text{m}.$$

As an example, a molecule with a net displacement of $0.2|e^-|$ through a distance of 0.1 nm has a dipole moment of 3.2×10^{-30} C $\cdot$ m or about 1D.

Rather than calculating the dipole moment from the charge distribution throughout a molecule, it is sometimes useful to characterize the dipole moment associated with various chemical bonds (bond dipoles). For reference, approximate values for permanent dipole moments associated with various bond dipoles are listed in Table 2.2.

Table 2.2   Permanent dipole bond moments
for representative chemical bonds.

Bond	Bond Dipole Moments (D)
H–C	0.30
C–N	0.22
C–O	0.86
C–I	1.29
H–N	1.31
C–Br	1.48
C–F	1.51
H–O	1.53
C–Cl	1.56
C=O	2.40
C≡N (cyano)	3.60

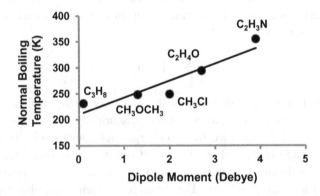

Fig. 2.11   The dependence of boiling point on molecular dipole moment for common solvents. The data points are for propane (C_3H_8), dimethyl ether (CH_3OCH_3), chloromethane (CH_3Cl), acetaldehyde (C_2H_4O), and acetonitrile (C_2H_3N).

As we shall see in the next chapter, molecular dipole moments are critically important for describing molecule–molecule interactions and they play an important role when selecting solvents. The near-field local electric fields generated by molecular dipoles are often sufficiently strong to remove atoms from compounds with high melting temperatures. As an example, common table salt (NaCl) is known to melt at 1074 K (800°C) yet it readily dissolves at room temperature in water, a polar solvent having a molecular dipole moment of 1.8 D.

Also, the boiling point of common polar solvents is correlated with increasing dipole moment as shown in Fig. 2.11, indicating that dipole–

dipole inter-molecular interactions play an important role in determining the physical properties of liquids.

2.6 Dipole Moments in External Electric Fields

Before ending this review, it is useful to mention a few of the electro-static consequences of a molecule with a **permanent** dipole moment. First, the electric field that develops around a dipole has interesting properties. Because the dipole has charges of equal but opposite polarity, the electric field *far* from the dipole will be small but in close proximity to the dipole, the electric fields can be large and highly non-uniform. This is schematically illustrated in Fig. 2.12 by plotting the dipolar electric field that develops when two charges $+q$, $-q$ are separated by a distance d. The magnitude of the dipole moment is given by

$$|p| \equiv p = |q|d. \tag{2.20}$$

The direction of p is defined from the negative to the positive charge.

It is also useful to consider the work required to rotate a dipole, fixed at a point in space but free to rotate about its midpoint when placed in

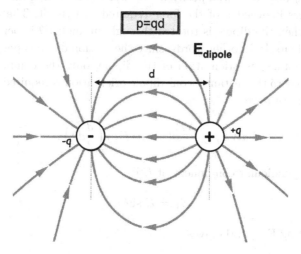

Fig. 2.12 An electric dipole is formed when two equal but opposite charges are displaced and rigidly held apart by a distance d. Such a charge configuration can be characterized by an electric dipole moment of magnitude $p = qd$. The direction of p points from negative to positive charge. The spatial dependence of the electric field that develops in close proximity to the dipole is schematically plotted.

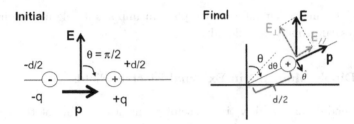

Fig. 2.13   Calculating the electrostatic potential energy of a freely rotating dipole in an applied external electric field $\vec{E}$ requires a calculation of the work to rotate the dipole to some angle θ. The initial dipole orientation (dipole perpendicular to E) is chosen because no work is required to slide the dipole from infinity into its final orientation in the presence of a uniform E-field.

a uniform applied electric field. We choose a coordinate system in which the electric field $\vec{E}$ is oriented at right angles to the initial position of the dipole as shown in Fig. 2.13. The calculation does not include the energy required to assemble the dipole. What is the work required to twist the dipole so it makes some angle θ with respect to the electric field $\vec{E}$?

Let's first focus on moving the positive charge. The work W_{+q} required to rotate $+q$ from its initial position ($\theta = \pi/2$) to some new angle specified by θ requires knowledge of the force required to twist it. The arc length traversed when the dipole is rotated through an angle $d\theta$ is perpendicular to the dipole itself. In other words, since the motion of $+q$ is perpendicular to the instantaneous orientation of the dipole, only the component of $\vec{E}$ perpendicular to the instantaneous dipole orientation is required. Referring to Fig. 2.12, we have

$$dW_{+q} = \vec{F} \cdot d\vec{l} = -(qE_\perp)\left(\frac{d}{2}d\theta\right). \qquad (2.21)$$

The perpendicular component of E is

$$E_\perp = E \sin\theta. \qquad (2.22)$$

Integrating Eq. (2.21) gives

$$W_{+q} = -\frac{1}{2}qd\,E\int_{\pi/2}^{\theta}\sin\theta d\theta = \frac{1}{2}pE\cos\theta. \qquad (2.23)$$

Similarly, the work required to rotate the $-q$ charge is given by

$$W_{-q} = \frac{1}{2}pE\cos\theta. \tag{2.24}$$

The total work is the sum

$$W_{\text{total}} = W_{+q} + W_{-q} = pE\cos\theta. \tag{2.25}$$

Once the work is calculated, $U(\theta)$, the electrostatic potential energy of a dipole oriented at an angle θ with respect to $\vec{E}$, is

$$U(\theta) \equiv -W_{\text{total}} = -pE\cos\theta = -\vec{p}\cdot\vec{E}. \tag{2.26}$$

The result in Eq. (2.26) for the electrostatic potential energy of a dipole in an external electric field is coordinate dependent. Equation (2.26) assumes the initial orientation of the dipole is specified by Fig. 2.12. Different initial orientations of the dipole (with respect to the applied electric field) can give rise to constant offsets from Eq. (2.26).

2.7 Chapter Summary

A general discussion of the interaction between atoms and molecules requires a basic understanding of classical electrostatic forces and the interaction between charged species. While isolated positive and negative ions interact via Coulomb's force law, electrically neutral molecules interact if they develop dipole moments due to small charge redistributions within the molecule itself. The redistribution of charge is influenced strongly by the electronegativity of the atoms comprising the molecule. Molecules with permanent dipole moments can interact even though they are electrically neutral. This interaction influences many physical properties like the boiling point of liquids.

2.8 Further Reading

Chapter Two References:

[feynman64] R.P. Feynman, R. Leighton, and M. Sands, The Feynman Lectures on Physics, V. II (Addison Wesley, 1964). pg. 10–8.

[kotz06] J.C. Kotz, P.M. Treichel and G.C. Weaver, *Chemistry and chemical reactivity*, 6th edition, Thomson Brooks/Cole, Belmont CA, USA (2006).

[mulliken34] R.S. Mulliken, J. Chem. Phys. **2**, 782–784 (1934).

[pauling32]   L. Pauling, J. Am. Chem. Soc. **54**, 3570–3582 (1932).

[vdwaals04]   J.D. van der Waals, *On the continuity of the gaseous and liquid states*, Dover Phoenix edition, Mineola NY (2004). A reprint of the seminal 1873 work of J. D. van der Waals, winner of the 1910 Nobel Prize in physics.

Further reading:

[atkins10]   P.W. Atkins and J. de Paula, *Physical Chemistry*, 9th edition, W.H. Freeman and Co., New York, NY, USA (2010).

[griffiths13]   D.J. Griffiths, Introduction to Electrodynamics, 4th ed., Pearson, (2013).

[israelachvili98]   J.N. Israelachvili, *Intermolecular and surface forces*. Academic Press, New York (1998).

[jeffrey 97]   G.A. Jeffrey, *An introduction to hydrogen bonding*. Oxford University Press (1997).

[pauling60]   L. Pauling, *The nature of the chemical bond*, 3rd ed., Cornell University Press, Ithaca (1960).

2.9   Problems

1. For N_2 gas, the van der Waals parameters are $a = 0.14\,\text{Pa}\,\text{m}^3/\text{mol}^2$ and $b = 3.9 \times 10^{-5}\,\text{m}^3/\text{mol}$. What is the pressure of 5 moles of nitrogen gas confined to a 1 liter container at 300 K? Compare your answer to that predicted using the ideal gas law.

2. Neon (Ne) and methane (CH_4) are two atomic scale objects that are similar to each other in many ways. Complete the following table

	Ne	CH_4
Total number of atoms		
Total number of electrons		
Atomic Weight (g)		
Boiling Point (K)		

You should find that neon is slightly heavier than methane, but its boiling temperature is about ∼80 K lower than CH_4. Why?

	vdW parameter a (units of Pa m^3/mol^2)	Boiling Temperature
Ar	0.13	87 K
H$_2$	0.025	20 K
He	0.0035	4 K
H$_2$O	0.554	373 K
Kr	0.23	166 K
N$_2$	0.14	77 K
Ne	0.021	27 K
NH$_3$	0.42	240 K
O$_2$	0.14	90 K
Xe	0.42	165 K

3. The van der Waals parameter a is a pressure correction term that accounts for attractive intermolecular interactions in a gas. You might expect that the value of a should therefore display a universal correlation with boiling temperature across a wide variety of different liquids. Use the data for representative gasses in the table and make a plot of boiling temperature vs. a. Is there any evidence for a correlation? If yes, what type of correlation do your observe?

4. A Mg^{+2} ion encounters a Cl^{-1} ion in a solvent with $\kappa = 5$ at a temperature of 300 K. Even though these two ions always feel a force of attraction for any separation distance, a simplistic model might require the electrostatic interaction energy to exceed the thermal energy (assumed to equal $k_B T$) before the motion of these two ions drives them toward each other to form a MgCl$_2$ precipitate. At what separation distance is this criterion met?

5. Sometimes the Lennard–Jones potential for two interacting molecules is written as

$$U(z) = \frac{A}{z^{12}} - \frac{B}{z^6}$$

How are the coefficients A, B related to the coefficients ε, σ in Eq. (2.15)?

6. Carbon is a remarkable element because it can share 1, 2 or 3 electrons forming single, double or triple bonds with another carbon atom. Does the presence of multiple bonds in a hydrocarbon molecule influence its interaction with other molecules of the same type? To investigate this question further, complete the following table. Does the presence

of single, double or triple bonds strongly influence the strength of inter-molecular interactions?

Molecular formula	C_2H_6	C_2H_4	C_2H_2
Molecule name	ethane	ethylene (ethene)	acetylene (ethyne)
Molecular structure			H–C≡C–H
Single, double or? triple bond			
Total number of atoms			
Total number of electrons			
Atomic Weight (g)			
Boiling Point (K)			
Polar or non-polar?			

7. A water molecule has an interior H–O–H angle of 104.5° with an OH length of 96 pm. The partial charge on each atom is calculated to be $a_H = +0.4$ and $a_O = -0.8$. From this information, calculate the dipole moment of a water molecule.

8. According to Table 2.2, the dipole moment of an O–H bond is 1.5 D. Using this information and the known geometric structure of H_2O, calculate the net dipole moment of water. How does your answer compare to that obtained in Problem 7?

9. The SO_2 molecule is bent and has a dipole moment of 1.60 D. The interior bond angle is 120°. The S–O bond length is 143 pm. What is the partial charge on each of the atoms?

10. Consider the 12-6 electrostatic potential energy function that is often used to characterize the interaction between two electrically neutral atoms separated by a distance z

$$U(z) = 4\varepsilon \left\{ \left(\frac{\sigma}{z}\right)^{12} - \left(\frac{\sigma}{z}\right)^6 \right\}$$

The function $U(z)$ has two adjustable parameters (σ and ε) that must be fit using experimental constraints. Suppose the equilibrium separation distance between the two atoms is measured to be z_o. How is σ related to z_o? What is the interaction potential energy in terms of ε when $z = z_o$? This quantity is roughly equivalent to the binding energy of the system, a quantity that is accessible from experimental data.

11. It is a useful exercise to develop a simple interaction potential energy model for an ionic bond that forms as a function of the separation distance z between two ions of opposite polarity (like Na^+ and Cl^-). One such model is specified by

$$U(z) = \frac{A}{z^p} - \frac{1}{4\pi\varepsilon_o}\frac{e^2}{z}.$$

The function $U(z)$ has two adjustable parameters, A and p, that can be adjusted to match experimental constraints.

(i) For a fixed value of p, how is A related to z_o, the equilibrium separation between the two ions?

(ii) What is the value of the interaction potential energy when $z = z_o$?

(iii) Calculate A for the case of NaCl. Assume $p = 12$. Assume the equilibrium spacing (z_o) for Na^+ and Cl^- ions is about $\sim$280 pm.

(iv) Make a plot of $U(z)$. How does the depth of the well compare to the measured binding energy of NaCl, which is about 8 eV/molecule?

12. Compare the similarities and differences between the short-range (Lennard–Jones) and long-range (ionic) interactions discussed in Problems 10 and 11 above. Make a plot to compare the 12-6 potential with the potential for an ionic bond when $p = 12$. To facilitate the comparison, make sure the depth of the minimum in each model is the same.

You will first need to calculate the value of A in Problem 11. For convenience, use $z_o = 280$ pm. Also, choose σ (Problem 10) to produce a minimum in the 12-6 potential when $z_o = 280$ pm. Choose ε (Problem 10) to match the depth of the well found in Problem 11 for $z_o = 280$ pm. In your plot, let z range from about 100 pm to about 1.5 nm. Write a few sentences to compare and contrast the two model interaction potentials.

13. Using only the chart in Fig. 2.7, rank the following chemical bonds in terms of their dipole moment, smallest to largest.

$$C\!-\!H,\ H\!-\!H,\ F\!-\!H,\ N\!-\!H,\ O\!-\!H$$

14. A very simple electrostatic model for a chemical covalent bond between two atoms is given below.

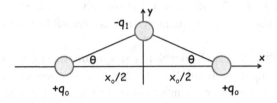

The two charges $+q_o$ are fixed at a distance x_o apart as shown. These two charges might represent the positive charge on the nucleus of each atom. A negative charge q_1 might represent electrons that (in this simple model) are constrained to some position $(0,y)$ along the perpendicular bisector as shown in the diagram. The angle θ is constrained such that $-\pi/2 < \theta < \pi/2$. In what follows, assume that the charges q_0 are stuck down as shown above.

(i) By inspection, write down the electrostatic potential energy U for the charge configuration drawn above?

(ii) Rewrite the answer to (i) in terms of θ?

(iii) In this simple model, the range of values for the ratio of q_1/q_o has not been specified. What restriction must be placed on q_1/q_o for U to have a negative value (i.e., for the charge configuration to have a lower energy)?

(iv) For a given q_o, there are many values of q_1 and θ that can satisfy the criterion found in (iii). Suppose we want to identify the minimum possible value for q_1 such that $U < 0$? What is the value of θ for the system to have $U < 0$ with a minimum value for q_1? The placement of charge for this minimum energy configuration has many similarities to a bonding orbital calculated using quantum mechanics.

15. The positions of the atoms and the fractional charges on each atom $\delta q_i = a_i|e^-|$ for an NH_3 molecule are given in the table. From this information, calculate the dipole moment of an NH_3 molecule.

Atom	x (in nm)	y (in nm)	z (in nm)	a
N	0	0	0	−0.8022
H	0	−0.09377	−0.03816	0.2674
H	0.08121	0.04689	−0.03816	0.2674
H	−0.08121	0.04689	−0.03816	0.2674

Chapter 3

Simple Models for Molecule–Molecule Interactions

It is often the practice in the discussion of classical tip–substrate forces to cite relevant equations obtained using assumptions that are not clearly understood because the derivation is never discussed in any detail. Such an approach seems to be the norm in today's literature rather than the exception. Students from some disciplines do not well tolerate the mathematics required. Other disciplines of a more applied bent seem in a great rush to move forward, blindly accepting what is evidently already known. While acceptable up to a point, the price paid is a loss of insight and intuition. If you are satisfied with this approach, then you can safely skip this chapter and continue to refer to the standard results for the van der Waals interaction which are summarized in Sec. 3.3.

In the following chapter, we attempt to systematically explain the essential points needed to understand the simple models of molecular interactions and we systematically review the physics underlying the hierarchy of interactions that lead to tip–substrate forces. We follow a pedagogical course, much in the spirit of Israelachvili's book on surface forces, in which many simple models are first developed and discussed before they are used to reach the next level of sophistication [israelachvili98].

While an effort is made to derive all the important results, the topics covered in this chapter are most accessible to students who have completed upper division courses in Electricity and Magnetism, Statistical Thermodynamics and Introductory Quantum Mechanics.

3.1   The Interaction of an Ion with a Dipole

While the force of interaction between two point charges (Sec. 2.2) is known by all who attend lectures in any introductory level physics class, the interaction between a point charge (ion) and a molecule is more interesting. By representing the molecule electrically as an electric dipole, the topic becomes tractable by recalling discussions often found in intermediate courses on electricity and magnetism. A discussion of this interaction forms an important first step toward understanding the tip–substrate interaction in AFM.

A review of different electrostatic interactions with increasing complexity is required to better understand the important issues. An overview to the variety of possible ion-molecule interactions is sketched in Fig. 3.1.

3.1.1   *An ion interacting with a fixed polar molecule*

Conisder a point charge Q positioned a distance z away from a molecule with a permanent dipole moment $p = qd$ situated in a uniform medium with dielectric constant κ. This situation approximates a dissolved ion in a solvent that interacts with a nearby molecule having a permanent dipole moment p as shown in Fig. 3.2. We use the notation z to designate the separation distance rather than the more common r in an attempt to develop a consistent notation that will carry through to our discussion of van der Waal's forces at the end of this chapter.

The dielectric constant of the surrounding medium is usually treated as a bulk quantity and book values of κ are used to estimate interaction

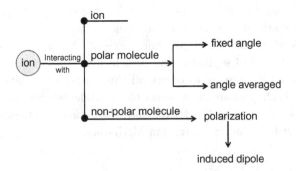

Fig. 3.1   An overview of ion–molecule interactions from the strongest (ion/ion) to the weakest (ion/non-polar).

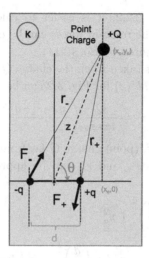

Fig. 3.2 A point charge Q placed a distance z away from an electric dipole $p = qd$. The interaction occurs in a surrounding homogenous solvent with a dielectric constant κ.

energies when the interacting entities are sufficiently far apart. As evident from Eq. (2.3), high values of κ for some liquids explains why they behave as good solvents — attractive Coulombic interactions between ions of opposite polarity are greatly reduced, thus allowing charged ions to remain dissolved rather than agglomerating to form a solid crystal.

It should be clear that using bulk (continuum) values for κ and treating κ as a constant under all circumstances is an approximation. The dielectric constant can be expected to decrease from its bulk value as the number of intervening solvent molecules decreases, a condition met when two interacting molecules/atoms come into close proximity to each other. The resulting interaction should *increase* since the dielectric constant is expected to *decrease* with separation, thereby enhancing the relevant electrostatic forces. These effects are not accounted for in the discussion that follows.

A model for the interaction between a point charge and a dipole requires the definition of a plane oriented to contain both these objects so that a relevant angle θ can be defined as shown in Fig. 3.2. The equations derived are coordinate specific to this definition of θ. At first glance, since the dipole is electrically neutral, you might expect no electrostatic interaction. However, each charge in the dipole separately experiences an interaction force with the point charge Q. The resulting forces are nearly equal and opposite as shown by F_+ and F_- in Fig. 3.2. A determination of the net interaction force requires the vector summation of F_+ and F_-.

As discussed in Sec. 2.2.2, it is easier to calculate the electrostatic potential energy $U(z)$. The relevant forces can be calculated by taking the derivative of $U(z)$ as required by Eq. (2.11).

If the zero of energy is taken when the charge Q is located very far from the dipole, by inspection, $U(z)$ for Fig. 3.2 can be written as

$$U(z) = \frac{Q}{4\pi\kappa\varepsilon_0}\left(\frac{q}{r_+} + \frac{(-q)}{r_-}\right). \tag{3.1}$$

When $r_+ > d$, $r_- > d$, (point charge Q far from the dipole), the square of the relevant distances r_+ and r_- can be approximated to first order in d by

$$r_-^2 = \left(x_o + \frac{d}{2}\right)^2 + y_o^2$$

$$= (x_o^2 + y_o^2) + x_o d + \left(\frac{d}{2}\right)^2 \simeq z^2\left(1 + \frac{x_o d}{z^2}\right) + \cdots \tag{3.2a}$$

$$r_+^2 = \left(x_o - \frac{d}{2}\right)^2 + y_o^2 \simeq z^2\left(1 - \frac{x_o d}{z^2}\right) + \cdots, \tag{3.2b}$$

where $z^2 \equiv x_o^2 + y_o^2$, and the assumption that $\left(\frac{d}{2}\right)^2 \ll z^2$ has been made.

This assumption allows a Taylor series expansion when evaluating $U(z)$

$$U(z) = \frac{Qq}{4\pi\kappa\varepsilon_0}\left[\left(z^2\left(1 - \frac{x_o d}{z^2}\right)\right)^{-\frac{1}{2}} - \left(z^2\left(1 + \frac{x_o d}{z^2}\right)\right)^{-\frac{1}{2}}\right]$$

$$= \frac{Qq}{4\pi\kappa\varepsilon_0}\frac{1}{z}\left[\left(1 + \frac{1}{2}\frac{x_o d}{z^2} + \cdots\right) - \left(1 - \frac{1}{2}\frac{x_o d}{z^2} + \cdots\right)\right]$$

$$\simeq \frac{Qq}{4\pi\kappa\varepsilon_0}\frac{1}{z}\frac{x_o d}{z^2} = \frac{Qp}{4\pi\kappa\varepsilon_0}\frac{1}{z^2}\cos\theta, \tag{3.3}$$

where the angle θ (in radians) measures the relative orientation between the point charge Q and the dipole with dipole moment $p = qd$ and $\cos\theta = x/d$ from Fig. 3.2. Note that $U(z)$ is really $U(z,\theta)$ and that it changes sign depending on θ:

$$U(z,\theta) > 0 \quad \text{when } 0 \le \theta \le \frac{\pi}{2},$$

$$\tag{3.4}$$

$$U(z,\theta) < 0 \quad \text{when } \frac{\pi}{2} \le \theta \le \pi.$$

Also note that $U(z,\theta)$ varies as z^{-2}. Recall that for two point charges, the interaction potential varied as z^{-1} (see Eq. (2.14)). Note also that the electrostatic interaction potential energy (see Eq. (3.3)) does not contain the electrostatic energy required to assemble the dipole located at the origin. This is a constant offset not included in Eq. (3.3) (or in any of the discussions that follow).

The signed magnitude of the force between the point charge and dipole (fixed θ) can always be obtained by applying Eq. (2.11)

$$|\vec{F}(z,\theta)| \simeq -\frac{\partial}{\partial z}\left[\frac{Qp}{4\pi\kappa\varepsilon_0}\frac{1}{z^2}\cos\theta\right] = \frac{2Qp}{4\pi\kappa\varepsilon_0}\frac{1}{z^3}\cos\theta. \tag{3.5}$$

The direction of the force is given by the sign associated with F — if $|\vec{F}(z,\theta)| > 0$, then the force is repulsive while if $|\vec{F}(z,\theta)| < 0$, the force is attractive. Clearly depending on θ, Eq. (3.5) can be either attractive or repulsive in nature.

3.1.2 An ion interacting with a polar molecule free to rotate

The situation discussed above approximates the electrostatic interaction between an ion and a polar molecule when the molecular dipole moment is permanently fixed in space. Such a situation might arise if the molecule is bound to an inert surface in a fixed orientation. What happens if the relative angle between the dipole moment and the point charge varies? This situation might arise when the position of the point charge fluctuates due to thermal energy. Or, perhaps the orientation of the dipole moment fluctuates in space but the position of the point charge is constant. Can we account for these situations by appropriately calculating the "angle average" of the electrostatic interaction potential energy? This calculation is equivalent to varying the angle θ in Fig. 3.2 and asking what is the average value for $U(z,\theta)$. What new results might come from this averaging procedure?

The answer to this question relies on computational techniques that are usually discussed in courses on statistical thermodynamics. When using the word thermodynamics, it is usually a short-hand way of asking "what would nature do if left alone?" The answer to this question is generally that a system tends to its lowest energy state.

Perhaps it is best to first discuss a general problem to review the appropriate techniques. Let us try to answer the question "What is the most likely value you will measure for some quantity "y" in a system containing r particles that is in equilibrium with a reservoir at temperature T?" In principle, y can be any quantity and indeed, we will eventually substitute $U(z,\theta)$ for y.

Table 3.1   A representative list of all possible values for a physical quantity y, the corresponding energies E, and the weighting factors w for each state.

Some physical quantity, y	Energy, E	Multiplicity, w
y_1	E_1	w_1
y_2	E_2	w_2
y_3	E_3	w_3
...	...	...
y_{N-1}	E_{N-1}	w_{N-1}
y_N	E_N	w_N

To answer this question, the main task is to collect information about how the energy of the system is related to the precise value of y. Suppose the quantity y spans a countable number of possible energy states N, so it is possible to tabulate all values of y (Table 3.1) along with the corresponding energies E. It may be possible that one value of y is more heavily weighted than other possible values. To account for this possibility, we must also enumerate the multiplicity or weight (w) which specifies the number of ways each value of y can be obtained.

In principle, the energies could describe discrete states — like electrons confined to a nanometer-size region of space; or the energies could be continuous — like gas with different kinetic energies. How might r particles in the system be distributed among the N energy levels? This is a general question of considerable interest to all scientific disciplines.

Let the symbol $\langle y \rangle$ be used to define the **thermal average** (also called the Boltzmann average) of y. If P_i is the probability that each value of y will be obtained, then we have

$$\langle y \rangle = \sum_i y_i P(y_i). \tag{3.6}$$

We rely on statistical thermodynamics to estimate the probabilities $P(y_i)$. If y_i specifies a possible state of the system having a weight or degeneracy w_i with an energy E_i that can be realized when the system is in thermal equilibrium with a reservoir held at temperature T, the probability $P(y_i)$ of finding the value y_i is given by

$$P(y_i) = \frac{w_i e^{\frac{-E_i}{k_B T}}}{\sum_i w_i e^{\frac{-E_i}{k_B T}}}, \tag{3.7}$$

where the denominator is required for proper normalization.

The exponential factor $e^{\frac{-E}{k_B T}}$ is the Boltzmann weighting factor which results when thermodynamics maximizes the configurational multiplicity (entropy) for a system of r particles distributed among N energy levels in thermal equilibrium at some temperature T. The quantity k_B is Boltzmann's constant with the value $k_B = 1.38 \times 10^{-23}$ J/K. The denominator is known as the partition function in statistical thermodynamics textbooks. If the variable y is continuous, then the energies E_i are likely continuous and the sums in Eqs. (3.6) and (3.7) must be replaced by an integral.

Estimating the weighting factors w_i for each state of the system becomes an important issue that depends on the details of the system under consideration. For the specific problem of a point charge interacting with a dipole, the proper weighting factor w comes from enumerating all possible ways the charge Q can be oriented with respect to the dipole (see Fig. 3.3). As shown in Fig. 3.3, Q is free to rotate in 3-dimensions, producing a strip of width $1/2pd\theta$ for a constant orientation angle θ.

Let $w(\theta)$ equal the number of ways that $\vec{p}$ and the point charge Q can be arranged between angles $(\theta - d\theta/2)$ and $(\theta + d\theta/2)$ to give the **same** interaction energy $U(z, \theta)$. We can estimate $w(\theta)$ by calculating the ratio of the area of the shaded strip in Fig. 3.3 (which includes all possible relative

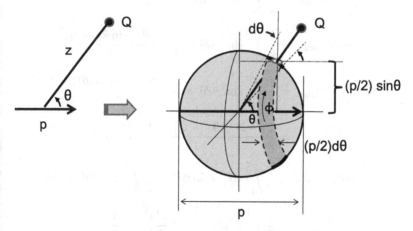

Fig. 3.3　When a point charge Q, a distance z from a dipole, is free to move with respect to the dipole, a calculation of the factor $w(\theta)$ for the dipole-point charge interaction $U(z, \theta)$ requires a proper weighting for each angle θ that gives the same interaction energy. This is equivalent to integrating the position of the charge Q around the azimuthal angle ϕ for each possible θ.

orientations for a fixed θ) to the area of the entire sphere

$$w(\theta) \equiv \frac{\text{area of shaded stripe}}{\text{surface area of sphere}} = \frac{\int_{\varphi=0}^{2\pi} d\varphi \times \left(\frac{D}{2}\sin\theta\right) \times \left(\frac{D}{2}d\theta\right)}{\int_{\varphi=0}^{2\pi} d\varphi \times \int_{\theta=0}^{\pi} \left(\frac{D}{2}\sin\theta\right) \times \left(\frac{D}{2}d\theta\right)}$$

$$= \frac{\sin\theta d\theta}{\int_0^\pi \sin\theta d\theta} = \frac{\sin\theta d\theta}{-\cos\theta|_0^\pi} = \frac{1}{2}\sin\theta d\theta \qquad (0 \leq \theta \leq \pi). \qquad (3.8)$$

This procedure for defining $w(\theta)$ can be used to define the angle averaged value for any continuous function $f(\theta)$ by

$$\langle f(\theta) \rangle |_{\text{angle}} = \int_0^\pi f(\theta) w(\theta) d\theta = \frac{1}{2}\int_0^\pi f(\theta) \sin\theta d\theta. \qquad (3.9)$$

Example 3.1: Calculate the angle weighted value of a constant C and various trigonometric functions $f(\theta) = \cos\theta, \sin\theta, \cos^2\theta, \sin^2\theta$, and $\sin\theta\cos\theta$.

It follows from the above discussion that we must evaluate Eq. (3.9) for the various functions listed.

$$\langle C \rangle |_{\text{angle}} = \frac{1}{2}\int_0^\pi C\sin\theta d\theta = -\frac{C}{2}\cos\theta\Big|_0^\pi = C$$

$$\langle \cos\theta \rangle |_{\text{angle}} = \frac{1}{2}\int_0^\pi \cos\theta\sin\theta d\theta = 0$$

$$\langle \sin\theta \rangle |_{\text{angle}} = \frac{1}{2}\int_0^\pi \sin^2\theta d\theta = \frac{1}{2} \times \frac{(\theta - \sin\theta\cos\theta)}{2}\Big|_0^\pi = \frac{\pi}{4}$$

$$\langle \sin^2\theta \rangle |_{\text{angle}} = \frac{1}{2}\int_0^\pi \sin^3\theta d\theta = \frac{1}{2}\frac{(\cos 3\theta - 9\cos\theta)}{12}\Big|_0^\pi$$

$$= \frac{1}{2}\frac{[(-1) - 9\cdot(-1)] - (1 - 9)}{12} = \frac{2}{3}$$

$$\langle \cos^2\theta \rangle |_{\text{angle}} = \frac{1}{2}\int_0^\pi \cos^2\theta\sin\theta d\theta = \frac{1}{2}\int_0^\pi \cos^2\theta d(\cos\theta)$$

$$= \frac{1}{2}\int_{-1}^1 x^2 dx = \frac{1}{2} \times \frac{x^3}{3}\Big|_{-1}^1 = \frac{1}{3}$$

$$\langle \sin\theta\cdot\cos\theta \rangle |_{\text{angle}} = \frac{1}{2}\int_0^\pi \cos\theta\sin^2\theta d\theta = \frac{1}{2}\frac{\sin^3\theta}{3}\Big|_0^\pi = 0$$

To calculate the angle averaged value of $U(z, \theta)$ in Eq. (3.5), you might be tempted to vary the angle θ defined in Fig. 3.3, following the procedure to weight each angle as laid out above. From Eq. (3.8), you would include the $\sin \theta$ weighting factor (the factor of $1/2$ can be dropped since it appears in both the numerator and denominator) and you would obtain the following

$$\langle U(z) \rangle |_{\text{angle}} \equiv \frac{\int_0^{2\pi} d\varphi \int_0^{\pi} U(z, \theta) \sin \theta d\theta}{\int_0^{2\pi} d\varphi \int_0^{\pi} \sin \theta d\theta},$$

where

$$U(z, \theta) = \frac{Qp}{4\pi \kappa \varepsilon_0} \frac{\cos(\theta)}{z^2} = U_o(z) f(\theta),$$

with

$$f(\theta) = \cos \theta \quad \text{and} \quad U_o(z) = \frac{Qp}{4\pi \kappa \varepsilon_0} \frac{1}{z^2} \tag{3.10}$$

$$\langle U(z) \rangle |_{\text{angle}} = \frac{U_o(z) \int_0^{\pi} \cos \theta \sin \theta d\theta}{\int_0^{\pi} \sin \theta d\theta}$$

$$= U_o(z) \cdot \left[\frac{-\frac{1}{2} \cos^2 \theta |_0^{\pi}}{- \cos \theta |_0^{\pi}} \right] = \frac{U_o(z)}{2} \cdot \frac{[(-1)^2 - (1)^2]}{(-1) - (1)} = 0.$$

The result is zero because the calculation treats all angles as equally likely. Instead, we must include a thermal average that additionally favors those angles with lowest energy. This requires the inclusion of the Boltzmann factor as given in Eq. (3.7) in addition to the angle weighting factor discussed above.

Following this approach we have

$$\langle U(z) \rangle \equiv \left\langle U(z, \theta) e^{\frac{-U(z, \theta)}{k_B T}} \right\rangle = \frac{\int_0^{2\pi} d\varphi \int_0^{\pi} U(z, \theta) e^{\frac{-U(z, \theta)}{k_B T}} \sin \theta d\theta}{\int_0^{2\pi} d\varphi \int_0^{\pi} e^{\frac{-U(z, \theta)}{k_B T}} \sin \theta d\theta}. \tag{3.11}$$

Since $U(z, \theta) = U_o(z) \cdot f(\theta)$ as before with $f(\theta) = \cos \theta$, we have

$$\langle U(z) \rangle = \frac{U_o(z) \int_0^{\pi} f(\theta) e^{-\left(\frac{U_o(z) f(\theta)}{k_B T} \right)} \sin \theta d\theta}{\int_0^{\pi} e^{-\left(\frac{U_o(z) f(\theta)}{k_B T} \right)} \sin \theta d\theta}.$$

Let

$$\beta = -\frac{U_o(z)}{k_B T},$$

then

$$\langle U(z) \rangle = U_o(z) \frac{\int_0^\pi f(\theta) e^{\beta f(\theta)} \sin\theta d\theta}{\int_0^\pi e^{\beta f(\theta)} \sin\theta d\theta}$$

$$= U_o(z) \frac{d}{d\beta} \ln \left(\int_0^\pi e^{\beta f(\theta)} \sin\theta d\theta \right). \qquad (3.12)$$

The last step follows from the identity

$$\frac{d}{d\beta} \ln \left(\int_0^\pi e^{\beta f(\theta)} \sin\theta d\theta \right) = \frac{1}{\left(\int_0^\pi e^{\beta f(\theta)} \sin\theta d\theta \right)} \times \left(\frac{\partial}{\partial\beta} \int_0^\pi e^{\beta f(\theta)} \sin\theta d\theta \right)$$

$$= \frac{1}{\left(\int_0^\pi e^{\beta f(\theta)} \sin\theta d\theta \right)} \times \left(\int_0^\pi f(\theta) e^{\beta f(\theta)} \sin\theta d\theta \right).$$

$$(3.13)$$

Equation (3.13) implies that only the natural logarithm of the integral must be evaluated (with $f(\theta) = \cos\theta$) before a derivative with respect to β is taken. This simplifies the discussion considerably since now a Taylor series expansion of $e^{\beta f(\theta)}$ is possible. If $\beta f(\theta) \ll 1$, we can write

$$\int_0^\pi e^{\beta f(\theta)} \sin\theta d\theta, \qquad (3.14)$$

$$\int_0^\pi e^{\beta f(\theta)} \sin\theta d\theta \simeq \int_0^\pi \left[1 + \beta f(\theta) + \frac{\beta^2}{2} f^2(\theta) + \cdots \right] \sin\theta d\theta$$

$$= \int_0^\pi \sin\theta d\theta + \beta \int_0^\pi f(\theta) \sin\theta d\theta$$

$$+ \frac{\beta^2}{2} \int_0^\pi f^2(\theta) \sin\theta d\theta + \cdots$$

$$= -\cos\theta|_0^\pi + 0 + \frac{\beta^2}{2} \int_0^\pi \cos^2\theta \sin\theta d\theta + \cdots$$

$$= 2 - \left(\frac{\beta^2}{2} \frac{1}{3} \right) \cos^3\theta|_o^\pi + \cdots = 2 + \frac{\beta^2}{3} + \cdots . \qquad (3.15)$$

This gives

$$\left\langle U(z,\theta) e^{-\frac{U(z,\theta)}{k_B T}} \right\rangle = U_o(z) \frac{d}{d\beta} \ln \left(\int_0^\pi e^{\beta f(\theta)} \sin\theta d\theta \right)$$

$$= U_o(z) \times \frac{d}{d\beta} \ln \left(2 + \frac{\beta^2}{3} + \cdots \right)$$

$$= U_o(z) \times \frac{1}{\left(2 + \frac{\beta^2}{3} + \cdots \right)} \times \frac{2\beta}{3}$$

$$= U_o(z) \times \frac{2\beta}{3} \times \frac{1}{2} \times \left(1 - \frac{\beta^2}{6} + \cdots\right)$$

$$= \frac{\beta}{3}U_o(z) + \cdots. \tag{3.16}$$

Since

$$\beta \equiv -\frac{U_o(z)}{k_BT} \quad \text{and} \quad U_o(z) = \frac{Qp}{4\pi\kappa\varepsilon_0}\frac{1}{z^2}, \tag{3.17}$$

we finally have the angle-averaged, thermal interaction energy between a point charge Q and a thermally rotating dipole p

$$\langle U(z) \rangle = -\frac{1}{3}\left(\frac{U_o(z)}{k_BT}\right) \times U_o(z) = -\frac{1}{3k_BT}\frac{(Qp)^2}{(4\pi\kappa\varepsilon_0)^2}\frac{1}{z^4}, \tag{3.18}$$

where the leading factor of $(3k_BT)^{-1}$ is a combination of the angle-weighted value of $\cos^2\theta = 1/3$ with the factor k_BT coming from the Boltzmann thermal factor which preferentially favors dipole orientations with lower energy states.

A comparison between the fixed angle point charge–dipole electrostatic interaction potential energy and the angle averaged thermal result in Eq. (3.18) is given in Fig. 3.4. In contrast to the fixed angle situation which

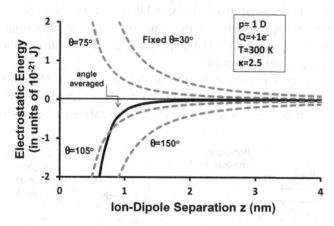

Fig. 3.4   A comparison between fixed angle (see Eq. (3.3)) and a thermal angle averaged (see Eq. (3.18)) calculation of the interaction energy $U(z,\theta)$ between a fixed dipole with dipole moment 1D and a moveable ion of charge $Q = +1.6 \times 10^{-19}$ C. For the fixed angle calculation, only a few representative angles of $30°, 75°, 105°$ and $150°$ are shown. The interaction can be attractive (negative) or repulsive (positive), depending on the dipole orientation. The thermal angle-averaged calculation $\langle U(z,\theta)\rangle$ is **always** attractive due to the thermal averaging introduced by the Boltzmann factor.

can be either repulsive (positive) or attractive (negative) depending on the location of the point charge, the thermal angle averaged result is **always** attractive. In addition, the interaction energy now varies as $1/z^4$ rather than $1/z^2$ for a fixed dipole-ion separation (see Eq. (3.3)). We conclude that the thermal angle averaged interaction is a shorter range interaction than the fixed angle result.

3.1.3 *Induction — the polarization of a non-polar molecule*

What happens when a non-polar atom or molecule comes into close proximity to a point charge? Lacking a dipole moment, you might expect no electrostatic interaction will take place. However it is possible to induce an electric dipole moment in the non-polar molecule, thereby insuring that it too can be attracted or repelled. The physics of this situation is outlined in Fig. 3.5.

If a non-polar molecule with a diameter of $\sim 2a_o$ finds itself in an applied electric field, the charge distribution can rearrange and in the process, an induced dipole moment p_{induced} can be formed. Experiment shows that to first order in E_{app}, the induced dipole is given by

$$\vec{p}_{\text{induced}} = \alpha \vec{E}_{\text{applied}}, \qquad (3.19)$$

where α is called the electronic polarizability and has units of $m \cdot C^2/N$. In principle, α depends on the frequency of $\vec{E}_{\text{app}}$. At this stage of the discussion, we will not include the frequency dependence; furthermore, what produces $\vec{E}_{\text{app}}$ is not important.

As shown in Fig. 3.6, a simple model to estimate the polarizability treats the molecule as a small positively charged nucleus (total charge $+q$) surrounded by a larger spherical negative charge distribution

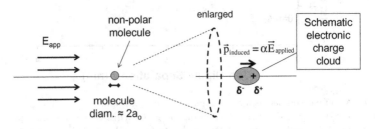

Fig. 3.5   A non-polar molecule with a diameter $\sim 2a_o$ in an external applied electric field E_{app}. The applied field distorts the electron distribution and the molecule becomes polarized, acquiring a net dipole moment p_{induced}. The proportionality constant between p_{induced} and E is the electronic polarizability α.

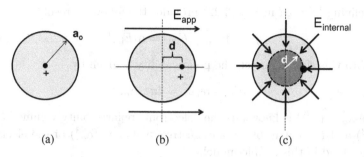

Fig. 3.6 A simple model to estimate the electronic polarizability α. In (a), an assumed spherically symmetric electron charge cloud with charge density ρ and radius a_o surrounds a positively charged nucleus. In (b), an external electric field E_{app} displaces the electronic charge cloud to the left, causing a net displacement d between the nucleus and the electron distribution. In (c), an internal electric field E_{internal} develops at the nucleus because of the displaced electronic charge. At the nucleus, E_{internal} opposes E_{app}.

(total charge $-q$) with radius a_o. The uniform electron charge density ρ (in C/m^3) is given by

$$\rho = \frac{-q}{\left(\frac{4}{3}\pi a_o{}^3\right)}. \tag{3.20}$$

In the presence of $\vec{E}_{\text{app}}$, the positive nucleus and the negative electronic charge cloud will be shifted in opposite directions, producing a net charge displacement d. The size of d will be determined by an equilibrium condition set by the internal electric field $\vec{E}_{\text{internal}}$ that develops because of the displaced electron cloud.

At the position of the nucleus, $\vec{E}_{\text{internal}}$ must balance $\vec{E}_{\text{app}}$ for a stable condition to result. If the electron cloud maintains its spherical shape, then the internal electric field (opposite in direction to the applied field at the positive nucleus) can be approximated using Gauss' law. Because of the assumed spherical symmetry,

$$|\vec{E}_{\text{internal}}| = \frac{1}{4\pi\varepsilon_o}\frac{q'}{d^2}, \tag{3.21}$$

where

$$q' = \frac{4\pi}{3}\rho d^3.$$

Using Eq. (3.20) for ρ gives

$$|\vec{E}_{\text{internal}}| = \frac{1}{4\pi\varepsilon_o}\frac{qd}{a_o^3}. \tag{3.22}$$

Defining $|p| = qd$ in Eq. (3.22) provides the following result

$$|p| = 4\pi\varepsilon_o a_o^3 |\vec{E}_{\text{int}}| = \alpha |\vec{E}_{\text{int}}|. \tag{3.23}$$

With this simple model, the polarizability is given by

$$\alpha = 4\pi\varepsilon_o a_o^3 = 4\pi\varepsilon_o \alpha_o, \tag{3.24}$$

where $\alpha_o = (a_o)^3$ is known as the "electronic polarizability volume" (units of m^3) and is comparable to an effective "volume" (a_o^3) of the electronic charge cloud in this simple model.

This model for polarizability is very simplistic and is only useful because it provides some physical insight into the microscopic origin of induced dipole moments. A more accurate model requires knowledge of the precise electron wave functions and application of perturbation theory in quantum mechanics.

Example 3.2:   Using Eq. (3.24), estimate the electronic polarizability for a H_2O molecule. Compare to the measured value of α for water.

The structure of a H_2O molecule is

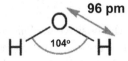

From this structure, the H-H distance is calculated to be around 150 pm. This suggests an estimate for the "radius" of a water molecule might lie between a lower limit of $150\,\text{pm}/2 = 75\,\text{pm}$ and an upper limit of $\sim$96 pm.

The polarizability can be estimated as
Lower limit:

$$\alpha = 4\pi\varepsilon_o\alpha_o = 4\pi \times 8.85 \times 10^{-12}\text{C}^2/\text{Nm}^2 \cdot (0.075 \times 10^{-9}\,\text{m})^3$$

$$= 4.7 \times 10^{-41}\,\text{m} \cdot \text{C}^2/\text{N}.$$

Upper limit:

$$\alpha = 4\pi\varepsilon_o\alpha_o = 4\pi \times 8.85 \times 10^{-12}\text{C}^2/\text{Nm}^2 \cdot (0.096 \times 10^{-9}\,\text{m})^3$$

$$= 9.8 \times 10^{-41}\,\text{m} \cdot \text{C}^2/\text{N}.$$

These estimates for α are low, giving a value that ranges between 28% to 59% of the book value for H_2O which is listed as $1.66 \times 10^{-40}\,\text{m} \cdot \text{C}^2/\text{N}$.

Table 3.2   A table of common solvents with a listing of their dielectric properties.

Material	Chemical Formula	Structure	K, dielectric constant[1]	p, (in Debye)[2,3]	$\alpha_o = \alpha/4\pi\epsilon_o$ (in $10^{-30}m^3$)
Acetone	$(CH_3)_2CO$		21	2.9	6.3
Isopropanel	$(CH_3)CHOH$		18	1.7	6.9
Ethanol	$(CH_3)_2CH_2OH$		24	1.7	5.1
Methanol	CH_3OH		33	1.7	3.3
Toluene	$C_6H_5CH_3$		2.4	0.4	12.3
Trichloro fluoromethane (CFC-11, Freon-11)	$CC\ell_3F$		2.0	0.4[5]	8.5
Water	H_2O		80	1.8	1.5

[1]Solvents with $\kappa < 15$ are generally considered non-polar.
[2]Tabulated dipole moments can be different than those listed, depending on whether they apply to the gaseous or liquid phase
[3]1 Debye = 3.33×10^{-30} C · m

Table 3.2 lists relevant dielectric constants, dipole moments and polarizability volumes for some common solvent molecules.

3.1.4   *The interaction of a point charge with a non-polar molecule*

If a non-polar molecule encounters the electric field produced by a point charge, then it is possible that an induced dipole is formed as shown in Fig. 3.7. In Fig. 3.7(a), a non-polar molecule is located in a dielectric

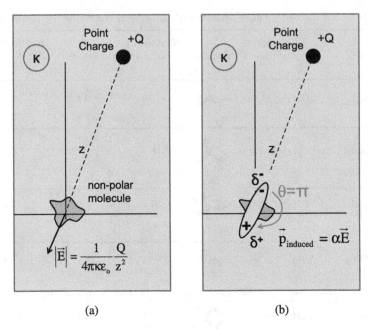

(a)   (b)

Fig. 3.7 In (a), a non-polar molecule interacts with a point charge separated by a distance z from the molecule. In (b), the electric field produced by the point charge at the molecule causes an induced dipole moment in the molecule that is always oriented as drawn.

medium having a dielectric constant κ. The non-polar molecule has no dipole moment as schematically illustrated. The point charge Q produces an electric field E that polarizes the molecule as shown in Fig. 3.6(b). The polarization is such that the angle between $\vec{p}_{\text{induced}}$ and $\vec{E}$ as defined in Fig. 3.7 is always 180° as shown. The electrostatic interaction potential energy that develops between the point charge and a dipole has already been derived (see Eq. (3.3)) and it must now be evaluated for $\theta = \pi$

$$U_{\text{induced}}(z) = \frac{Q\,p_{\text{induced}}}{4\pi\kappa\varepsilon_0}\frac{1}{z^2}\cos\theta\bigg|_{\theta=\pi} = -\frac{Q\,p_{\text{induced}}}{4\pi\kappa\varepsilon_0}\frac{1}{z^2} \qquad (3.25)$$

with $\vec{p}_{\text{induced}} = \alpha\vec{E}$, where $\vec{E}$ is the electric field generated by a point charge Q. We therefore have

$$U_{\text{induced}}(z) = -\frac{Q\,\alpha|\vec{E}|}{4\pi\kappa\varepsilon_0}\frac{1}{z^2} = -\frac{\alpha Q^2}{(4\pi\kappa\varepsilon_0)^2}\frac{1}{z^4}. \qquad (3.26)$$

Example 3.3: A Li ion is $2\,\text{nm}$ away from a non-polar molecule which is dissolved in a liquid with a dielectric constant of $\kappa = 2.3$. The molecule has a polarizability volume $\alpha_o = 4.8 \times 10^{-30}\,\text{m}^3$. Calculate (a) the electrostatic interaction potential energy (in eV) and (b) the force between the non-polar molecules and the Li ion.

(a) $\quad U_{\text{induced}}(z) = -\dfrac{\alpha Q^2}{(4\pi\kappa\varepsilon_0)^2}\dfrac{1}{z^4} = -\dfrac{4\pi\varepsilon_0\alpha_o Q^2}{(4\pi\kappa\varepsilon_0)^2}\dfrac{1}{z^4} = -\dfrac{\alpha_o Q^2}{(4\pi\varepsilon_0)}\dfrac{1}{\kappa^2}\dfrac{1}{z^4}$

$$= -\frac{(4.8 \times 10^{-30}\,\text{m}^3)(+1.6 \times 10^{-19}\,\text{C})^2}{4\pi \times 8.85 \times 10^{-12} \times \frac{\text{C}^2}{\text{Nm}^2}}$$

$$\times \frac{1}{(2.3)^2} \times \frac{1}{(2 \times 10^{-9}\,\text{m})^4}$$

$$= -1.105 \times 10^{-57} \times 1.89 \times 10^{-1} \times 6.25 \times 10^{34}\,\text{J}$$

$$= -1.31 \times 10^{-23}\,\text{J} \times \frac{1\text{eV}}{1.6 \times 10^{-19}\,\text{J}} = -8.2 \times 10^{-5}\,\text{eV}.$$

This interaction energy is ~ 300 times smaller than thermal energies which are typically estimated as $k_B T = 4.21 \times 10^{-21}\,\text{J} = 0.026\,\text{eV}$ at $T = 300\,\text{K}$. In general, the ion and molecule will bind or "condense" when their interaction energy is large compared to thermal energies. Similarly, we might expect the ions will remain dispersed in solution until the interaction energy becomes comparable to $k_B T$.

(b) $\quad F = -\dfrac{\partial U_{\text{induced}}(z)}{\partial z} = (-4)\dfrac{\alpha Q^2}{(4\pi\kappa\varepsilon_0)^2}\dfrac{1}{z^5}$

$$= -4 \times \frac{4\pi\varepsilon_0\alpha_o Q^2}{(4\pi\kappa\varepsilon_0)^2}\frac{1}{z^5} = -\frac{4\alpha_o Q^2}{(4\pi\varepsilon_0)}\frac{1}{\kappa^2}\frac{1}{z^5}$$

$$= -\frac{4(4.8 \times 10^{-30}\,\text{m}^3)(+1.6 \times 10^{-19}\,\text{C})^2}{4\pi \times 8.85 \times 10^{-12}\frac{\text{C}^2}{\text{Nm}^2}}$$

$$\times \frac{1}{(2.3)^2} \times \frac{1}{(2 \times 10^{-9}\,\text{m})^5}$$

$$= -4.42 \times 10^{-57} \times 1.89 \times 10^{-1} \times 3.13 \times 10^{43}\quad\text{N}$$

$$= -2.6 \times 10^{-14}\,\text{N} \quad\text{(attractive)}.$$

This force is small, about 350 times smaller than the smallest force ($\sim 10\,\text{pN} = 10 \times 10^{-12}\,\text{N}$) that can be measured using an AFM.

The important message from this discussion is that ion-dipole interactions vary as z^{-4}, irrespective of whether the dipole is a permanent
characteristic of the molecule (see Eq. (3.18)) or induced by an external
electric field (see Eq. (3.26)).

3.2   Molecule-Molecule Interactions

Within the context of AFM, it is useful to have models that account for
the force that a tip of radius R experiences when positioned a distance d
above a flat sample as shown in Fig. 3.8. Any net force must result from
interactions between atoms that comprise the tip and sample. Usually, the
atoms in the tip and sample are electrically neutral, so ion-ion or ion-
molecule interactions are not often relevant. To understand the nature of
the tip–substrate interaction, it therefore becomes important to consider
the interaction of neutral atoms (or molecules) with other neutral atoms
(or molecules).

A number of distinct possibilities arise as shown in Fig. 3.9 and these
situations, when systematically developed, lead to the classic expressions
for the Keesom, Debye and London interactions. All three of these interactions can be further grouped into a single category called a van der Waals
interaction.

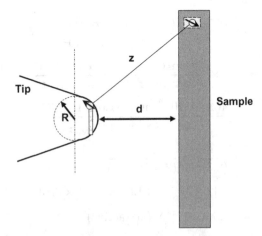

Fig. 3.8   The ultimate goal is to use our understanding of molecule-molecule forces to
estimate the force between a sharp tip of radius R positioned a distance d from a flat
substrate.

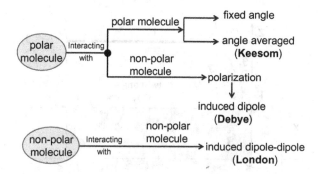

Fig. 3.9   An overview of molecule–molecule van der Waal interactions ranging from the strongest (polar/polar) to the weakest (non-polar/non-polar).

The important point that will develop is that electrically neutral atoms/molecules can electrostatically interact with each other. In what follows, we focus on how these interactions vary with separation distance, since it is this quantity we control with an AFM.

3.2.1   The interaction of two polar molecules

Consider two polar molecules with permanent dipole moments $\vec{p}_1, \vec{p}_2$, embedded in a continuous dielectric with dielectric constant κ and separated by a fixed distance r as shown in Fig. 3.10. The two molecules lie in the z-y plane and can be rotated through arbitrary angles ϑ_1, ϑ_2 with respect to the z-axis as shown. The standard definitions for the spherical polar coordinates (r, θ, φ) are assumed. In addition, the orientation of dipole $\vec{p}_2$ can be twisted with respect to $\vec{p}_1$ through an angle ζ in the x-z plane as shown. The two dipoles will interact electrostatically because the dipole $\vec{p}_1$ produces an electric field that exerts a force on the charges in dipole $\vec{p}_2$.

The interaction potential energy will now depend on three angles $(\vartheta_1, \vartheta_2, \zeta)$ that specify the relative orientation of the two dipoles. There is no standard definition of the angles ϑ_1, ϑ_2, so the form of any analytical result is tied to the way these angles are defined.

To calculate the electrostatic interaction potential energy, we can make use of the result given by Eq. (2.26)

$$U(r, \vartheta_1, \vartheta_2, \zeta) = -\vec{p}_2 \cdot \vec{E}_1,$$

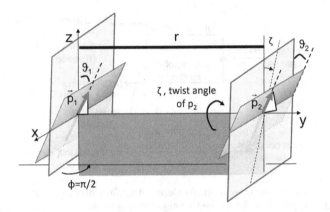

Fig. 3.10   The geometry required to analyze the interaction of two dipoles. The dipole p_1 is tilted by an angle ϑ_1 in z–y plane. The dipole p_2 is tilted by an angle ϑ_2 in the y–z plane. In addition, p_2 is twisted by an angle ζ in the x–z plane. The dipole-dipole separation distance is r.

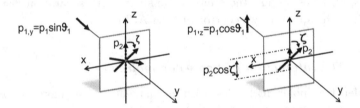

Fig. 3.11   The interaction potential energy between two permanent dipoles will contain two contributions that depend on the components of p_1. In general, the rotation angle of p_2 is ζ as shown.

where

$$\vec{E}_1(\vec{r}, \theta) = \frac{|\vec{p}_1|}{4\pi\kappa\varepsilon_o}\frac{1}{r^3}(2\cos\theta\hat{r} + \sin\theta\hat{\theta}). \qquad (3.27)$$

Here $\vec{E}_1$ is the electric field produced by dipole $\vec{p}_1$.

When evaluating $U(r, \vartheta_1, \vartheta_2, \zeta)$ in Eq. (3.27), it is useful to first consider the components of p_1 parallel to y ($p_{1,y}$) and z ($p_{1,z}$) as limiting cases.

These two geometries are sketched in Fig. 3.11. These limiting cases illustrate that $U(r, \vartheta_1 = \pi/2, \vartheta_2, \zeta)$ cannot depend on ζ since p_2 is always perpendicular to p_1 for any value of ζ. Similarly, $U(r, \vartheta_1 = 0, \vartheta_2, \zeta)$ must depend on the projection of p_2 along the z-axis and hence must depend on $\cos\zeta$. These two results are useful to understand the two contributions to $U(r, \vartheta_1, \vartheta_2, \zeta)$ for any arbitrary value of ϑ_1.

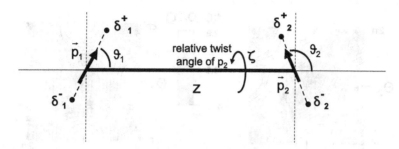

Fig. 3.12 Defining the coordinates and angles for the dipole–dipole interaction between two dipoles separated by a distance z.

The final expression for $U(r, \vartheta_1, \vartheta_2, \zeta)$ includes two terms that represent the relative contributions for the two components of $\vec{p}_1$ shown in Fig. 3.11. A derivation of the final result requires some geometrical dexterity and uses the standard approximation for the electric field $\vec{E}_1$ (Eq. (3.27)) from a "pure" or "point" dipole in which $r \gg d$, where d is the separation between the charges comprising the dipole and r represents the dipole–dipole separation.

After completing the calculation in a standard spherical-polar coordinate system, it is convenient to re-label the separation distance r as z and to redefine the angles as shown in Fig. 3.12.

The final result for the dipole–dipole interaction energy is

$$U_{\text{total}}(z, \vartheta_1, \vartheta_2, \zeta) = -\frac{1}{4\pi\kappa\varepsilon_o}\frac{p_1 p_2}{z^3}(2\cos\vartheta_1\cos\vartheta_2 - \sin\vartheta_1\sin\vartheta_2\cos\zeta)$$

$$= \frac{p_1 p_2}{4\pi\kappa\varepsilon_o}\frac{1}{z^3} \times f(\vartheta_1, \vartheta_2, \zeta), \tag{3.28}$$

where

$$f(\vartheta_1, \vartheta_2, \zeta) \equiv -(2\cos\vartheta_1\cos\vartheta_2 - \sin\vartheta_1\sin\vartheta_2\cos\zeta). \tag{3.29}$$

The angular function $f(\vartheta_1, \vartheta_2, \zeta)$ is bounded such that

$$-2 \leq f(\vartheta_1, \vartheta_2, \zeta) \leq 2. \tag{3.30}$$

Plots of $f(\vartheta_1, \vartheta_2, \zeta)$ are useful to show the angular dependence of the interaction energy. Two special cases are provided in Fig. 3.13.

The relative orientation of two dipoles is sketched in Fig. 3.14 for a few angles to show the configuration producing the minimum (attractive) and maximum (repulsive) interaction. These orientations are also indicated in Fig. 3.13 on the $\zeta = 0$ plot. These alignments provide a quick reference (in conjunction with Eq. (3.28)) to estimate the interaction energy

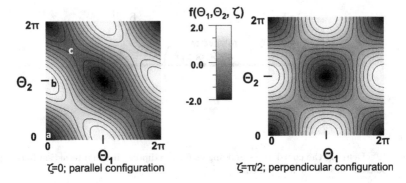

Fig. 3.13   Representative contour plots of $f(\vartheta_1, \vartheta_2, \zeta)$ (Eq. (3.29)) for $\zeta = 0$ and $\zeta = \pi/2$. The labels a, b and c on the plot for $\zeta = 0$ shows the locations for the three relative orientations of two dipoles sketched in Fig. 3.14.

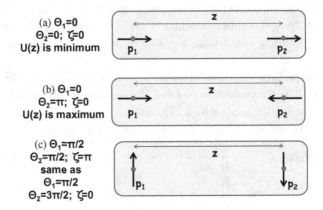

Fig. 3.14   Three representative configurations for two dipoles separated by a distance z.

between two polar molecules that have a *fixed orientation* with respect to each other. Such a situation might arise if, for instance, two polar molecules are chemically bound to a planar surface.

3.2.2   *The angle-averaged interaction between two polar molecules*

If the orientation of the dipoles p_1 and p_2 can vary due to thermal effects, a weighted angle average must be calculated in order to estimate the net interaction energy. Following a similar procedure already outlined in

Sec. 3.1.2 above, we define $\langle U(z) \rangle$ as

$$\langle U(z) \rangle \equiv \left\langle U(z, \Theta_1, \Theta_2, \zeta) e^{\frac{-U(z, \Theta_1, \Theta_2, \zeta)}{k_B T}} \right\rangle, \tag{3.31}$$

where now

$$U(z, \Theta_1, \Theta_2, \zeta) = -\frac{p_1 p_2}{4\pi\kappa\varepsilon_o} \frac{1}{z^3} [2 \cos\Theta_1 \cos\Theta_2 - \sin\Theta_1 \sin\Theta_2 \cos\zeta]. \tag{3.32}$$

Here $\langle \ \rangle$ again represents the thermal average over all angles and implies that the two objects are free to move by thermal motion as they interact. Let

$$U(z, \Theta_1, \Theta_2, \zeta) = U_o(z) \cdot f(\Omega), \tag{3.33}$$

where

$$U_o(z) = \frac{p_1 p_2}{4\pi\kappa\varepsilon_0} \frac{1}{z^3} \tag{3.34}$$

and

$$f(\Omega) = -2 \cos\Theta_1 \cos\Theta_2 + \sin\Theta_1 \sin\Theta_2 \cos\zeta. \tag{3.35}$$

To find the angle averaged value for $U(z, \Theta_1, \Theta_2, \zeta)$, we then must evaluate

$$\langle U(z) \rangle \equiv \frac{\int_0^{2\pi} d\zeta \int_{\Theta_1=0}^{\pi} \int_{\Theta_2=0}^{\pi} U_o(z) f(\Omega) e^{\frac{-U_o(z) f(\Omega)}{k_B T}} \sin\Theta_1 d\Theta_1 \sin\Theta_2 d\Theta_2}{\int_0^{2\pi} d\zeta \int_{\Theta_1=0}^{\pi} \int_{\Theta_2=0}^{\pi} e^{\frac{-U_o(z) f(\Omega)}{k_B T}} \sin\Theta_1 d\Theta_1 \sin\Theta_2 d\Theta_2}. \tag{3.36}$$

The discussion below follows closely that found in Sec. 3.1.2 for the case of a point charge near a dipole. Defining

$$\beta = -\frac{U_o(z)}{k_B T} \tag{3.37}$$

and realizing that

$$U_o(z) \times \frac{d}{d\beta} \ln \left(\int_0^{2\pi} d\zeta \int_{\Theta_1=0}^{\pi} \int_{\Theta_2=0}^{\pi} e^{\beta f(\Omega)} \sin\Theta_1 d\Theta_1 \sin\Theta_2 d\Theta_2 \right)$$

$$= \frac{U_o(z) \int_{\zeta=0}^{2\pi} d\varphi \int_{\Theta_1=0}^{\pi} \int_{\Theta_2=0}^{\pi} f(\Omega) e^{\beta f(\Omega)} \sin\Theta_1 d\Theta_1 \sin\Theta_2 d\Theta_2}{\left(\int_{\zeta=0}^{2\pi} d\varphi \int_{\Theta_1=0}^{\pi} \int_{\Theta_2=0}^{\pi} e^{\beta f(\Omega)} \sin\Theta_1 d\Theta_1 \sin\Theta_2 d\Theta_2 \right)} \tag{3.38}$$

allow us to focus on the integral

$$I = \int_0^{2\pi} d\zeta \int_{\Theta_1=0}^{\pi} \int_{\Theta_2=0}^{\pi} e^{\beta f(\Omega)} \sin\Theta_1 d\Theta_1 \sin\Theta_2 d\Theta_2$$

$$\simeq \int_0^{2\pi} d\zeta \int_{\Theta_1=0}^{\pi} \int_{\Theta_2=0}^{\pi} \left[1 + \beta f(\Omega) + \frac{\beta^2}{2} f^2(\Omega) + \cdots\right] \sin\Theta_1 d\Theta_1 \sin\Theta_2 d\Theta_2.$$

$$(3.39)$$

After some algebra, we find the above integral is given to order β^2 by

$$I = 8\pi \times \left(1 + \frac{\beta^2}{3} + \cdots\right) \tag{3.40}$$

$\langle U(z) \rangle$ can now be evaluated by taking a logarithmic derivative

$$\langle U(z) \rangle \equiv \left\langle U(z, \Theta_1, \Theta_2, \zeta) e^{-\frac{U(z,\Theta_1,\Theta_2,\zeta)}{k_B T}} \right\rangle = U_o(z) \frac{d}{d\beta} \ln(I)$$

$$= U_o(z) \frac{d}{d\beta} \left[\ln(8\pi) + \ln\left(1 + \frac{\beta^2}{3} + \cdots\right)\right]$$

$$= 0 + U_o(z) \frac{1}{1 + \frac{\beta^2}{3}} \times \frac{2\beta}{3}$$

$$\simeq U_o(z) \times \frac{2\beta}{3} \times \left(1 - \frac{\beta^2}{3} + \cdots\right) \approx U_o(z) \frac{2\beta}{3}. \tag{3.41}$$

Since

$$\beta = -\frac{U_o(z)}{k_B T} \quad \text{and} \quad U_o(z) = \frac{p_1 p_2}{4\pi\kappa\varepsilon_0} \frac{1}{z^3}, \tag{3.42}$$

the angle-averaged interaction energy between two dipoles separated by a distance z becomes

$$\langle U_{\text{Keesom}}(z) \rangle = -\frac{2}{3} \frac{1}{k_B T} \left(\frac{p_1 p_2}{4\pi\kappa\varepsilon_0}\right)^2 \frac{1}{z^6}. \tag{3.43}$$

Equation (3.43) was obtained by allowing both dipoles freedom to rotate since Eq. (3.38) integrates over both angles θ_1 and θ_2. This double counts equivalent configurations between the two dipoles. The correct answer is therefore obtained by multiplying Eq. (3.43) by a factor of $\frac{1}{2}$, giving

$$\langle U_{\text{Keesom}}(z) \rangle = -\frac{1}{3} \frac{1}{k_B T} \left(\frac{p_1 p_2}{4\pi\kappa\varepsilon_0}\right)^2 \frac{1}{z^6}. \tag{3.44}$$

Another way of obtaining this result is to lock the orientation of dipole p_1 such that $\Theta_1 = \pi/2$. Then $f(\Omega)$ in Eq. (3.36) simplifies to $\sin \Theta_2 \cos \zeta$. Under these circumstances, it is relatively straightforward to show that the 2/3 value of the factor found in Eq. (3.43) is reduced to a value of 1/3.

The interaction energy acquires a characteristic z^{-6} dependence on dipole–dipole separation and is inversely proportional to temperature. This result was derived in 1921 by W.H. Keeson. He showed that two molecules with permanent dipole moments undergoing thermal motion will on average give rise to an attractive intermolecular force[keesom21].

Example 3.4: Calculate the Keesom angle-averaged electrostatic potential energy for two polar molecules with dipole moments of 1.7 D that are separated by a distance of 2 nm in a medium having a dielectric constant of 24. Assume the temperature of the system is at 300 K.

$$\langle U_{\text{Keesom}}(z) \rangle = \frac{1}{3} \frac{1}{k_B T} \left(\frac{p_1 p_2}{4\pi k \varepsilon_0} \right)^2 \frac{1}{z^6}$$

$$= -\frac{1}{3} \left(\frac{1}{(1.38 \times 10^{-23} \text{J/K})(300\text{K})} \right)$$

$$\times \left(\frac{\left((1.7\text{D}) \times \left(\frac{3.33 \times 10^{-30} \text{cm}}{1\text{D}} \right) \right)^2}{4\pi \times (24) \times 8.85 \times 10^{-12} \frac{\text{C}^2}{\text{Nm}^2}} \right)^2 \frac{1}{(2 \times 10^{-9}\text{m})^6}$$

$$= -\frac{1}{3} \left(\frac{1}{1.24 \times 10^{-20}\text{J}} \right) \left(\frac{(5.66 \times 10^{-30})^2}{2.67 \times 10^{-9}} \text{Nm}^4 \right)^2$$

$$\times \frac{1}{6.40 \times 10^{-53}\text{m}^6}$$

$$= -\frac{1}{3}(8.06 \times 10^{19}\text{J}^{-1}) \times (1.44 \times 10^{-100}\text{N}^2\text{m}^8)$$

$$\times (1.56 \times 10^{52}\text{m}^{-6})$$

$$= -6.03 \times 10^{-29}\text{J} \times \left(\frac{1\,\text{eV}}{1.6 \times 10^{-19}\text{J}} \right)$$

$$= -3.8 \times 10^{-10}\text{eV}$$

(Continued)

Example 3.4: (*Continued*)

This is very small interaction energy when compared with thermal energies.

The force between the two dipoles is attractive and can be found by taking the negative gradient of $\langle U_{\text{Keesom}}(z) \rangle$ with respect to z (Eq. (2.11)). The force F is found to be

$$F = -\frac{\partial \langle U_{\text{Keesom}}(z) \rangle}{\partial z} = -\left[-\frac{1}{3}\frac{1}{k_B T}\left(\frac{p_1 p_2}{4\pi\kappa\varepsilon_0} \right)^2 \frac{(-6)}{z^7} \right]$$

$$= -6.03 \times 10^{-29}\,\text{J} \times \left(\frac{6}{2 \times 10^{-9}\text{m}} \right)$$

$$= -1.8 \times 10^{-19}\text{N} = -18 \text{ aN(attractive)}$$

3.2.3 *The interaction between a di-polar molecule and a non-polar molecule*

The electric field far from a polar molecule with a permanent dipole moment is complicated, changing in orientation and strength depending on the exact coordinates (r, θ) in a polar spherical coordinate system. Unlike an x–y coordinate system in which the unit vectors always point in the same direction, the unit vectors $(\hat{r}, \hat{\theta})$ in a spherical polar coordinate system change direction as a function of (r, θ). Consider a polar molecule with a permanent dipole moment p_1 embedded in a uniform dielectric with dielectric constant κ oriented along the z-axis, the far-field components (E_r, E_θ) of the dipole's electric field when $r \gg d$ (d = separation between dipole charges) is known from classical electrostatics[griffiths13].

$$\vec{E}_{\text{dipole}}(\vec{r}, \theta) = \frac{|\vec{p}_1|}{4\pi\kappa\varepsilon_o}\frac{1}{r^3}\left(2\cos\theta\hat{r} + \sin\theta\hat{\theta} \right). \tag{3.45}$$

The magnitude of the dipolar electric field is given by

$$|\vec{E}_{\text{dipole}}(\vec{r}, \theta)| = \frac{|\vec{p}_1|}{4\pi\kappa\varepsilon_o}\frac{1}{r^3}(4\cos^2\theta + \sin^2\theta)^{\frac{1}{2}}$$

$$= \frac{|\vec{p}_1|}{4\pi\kappa\varepsilon_o}\frac{1}{r^3}(3\cos^2\theta + 1)^{\frac{1}{2}}. \tag{3.46}$$

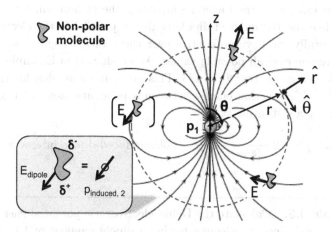

Fig. 3.15   Non-polar molecules located at a radial distance r in the far-field vicinity of a polar molecule with a permanent dipole moment p_1. Induced dipole moments $p_{induced,2}$ on the non-polar molecules align with the local direction of the electric field as shown in the inset.

At a fixed distance r from p_1, the magnitude of the electric field clearly depends on the angle θ.

If a second, non-polar molecule is located a distance r from the polar molecule, at first thought you might claim there will be no net interaction. However, the second molecule can be polarized by E_{dipole} as shown in Fig. 3.15. This case was considered by P. Debye in 1921. He showed that a net attractive force can exist between two molecules even if only one of them has a permanent dipole moment. As shown schematically in Fig. 3.15, the orientation of the induced dipole will depend on the precise direction of E_{dipole} at any point (r, θ).

If the second molecule has non-zero polarizability, then, according to Eq. (3.19) it will acquire a dipole moment in the direction of the dipolar electric field

$$\vec{p}_{induced,2} = \alpha_2 \vec{E}_{dipole}, \tag{3.47}$$

where α_2 is the polarizability of molecule 2. Since $\vec{p}_{induced,2}$ is always parallel to $\vec{E}_{dipole}$, the electrostatic interaction energy will then be given by

$$U(r, \theta) = -\vec{p}_{induced,2} \cdot \vec{E}_{dipole} = -\alpha_2 E_{dipole}^2$$

$$= -\alpha_2 \left(\frac{|\vec{p}_1|}{4\pi\kappa\varepsilon_o} \frac{(3\cos^2\theta + 1)^{\frac{1}{2}}}{r^3} \right)^2 = -\frac{p_1^2 \alpha_2}{(4\pi\kappa\varepsilon_o)^2} \frac{(3\cos^2\theta + 1)}{r^6}. \tag{3.48}$$

In this case, a thermal average involving the Boltzmann factor is not required because the induced dipole is always parallel to the electric field. Experimentally, we have no control over the position angle θ so we may as well average over all possible values. As we showed in Example 3.1, the weighted average of $\langle \cos^2 \theta \rangle = 1/3$. This gives an expression for the position averaged interaction energy between a polar and non-polar molecule separated by a distance $r = z$

$$U_{\text{Debye}}(z) = -2 \cdot \frac{p_1^2 \alpha_2}{(4\pi\kappa\varepsilon_o)^2} \frac{1}{z^6} \quad \text{(polar-induced-dipole interaction)}$$

$$(3.49)$$

Example 3.5: Calculate the Debye electrostatic potential energy for two identical dipolar molecules having a dipole moment of 1.7 D that are separated by a distance of 2 nm in a fluid having a uniform dielectric constant of 24. Take the polarizability volume for each molecule to be $\alpha_o = 5.1 \times 10^{-30}\,\text{m}^3$.

Since the two molecules are dipolar, they will experience a Keesom interaction in addition to a dipole/induced dipole interaction. The Debye formula (Eq. (3.49)) is used to describe the polar/non-polar interaction.

$$U_{\text{Debye}}(z = 2\,\text{nm})$$

$$= -2 \times \frac{p^2\alpha}{(4\pi\kappa\varepsilon_o)^2} \frac{1}{z^6} = -2 \times \frac{p^2 4\pi\varepsilon_o \alpha_o}{(4\pi\kappa\varepsilon_o)^2} \frac{1}{z^6} = -2 \frac{p^2 \alpha_o}{(4\pi\varepsilon_o)} \frac{1}{\kappa^2} \frac{1}{z^6}$$

$$= -2 \left[\frac{\left((1.7\,\text{D}) \times \left(\frac{3.33 \times 10^{-30}\,\text{Cm}}{1\,\text{D}} \right) \right)^2 (5.1 \times 10^{-30}\,\text{m}^3)}{4\pi \times 8.85 \times 10^{-12}\,\frac{\text{C}^2}{\text{Nm}^2}} \right]$$

$$\times \left(\frac{1}{24} \right)^2 \frac{1}{(2 \times 10^{-9}\,\text{m})^6}$$

$$= -2 \left[\frac{1.63 \times 10^{-88}\,\text{C}^2\text{m}^3}{1.11 \times 10^{-10}\,\frac{\text{C}^2}{\text{Nm}^2}} \right] \times (1.74 \times 10^{-3}) \times (1.56 \times 10^{52}\,\text{m}^{-6})$$

$$= -7.4 \times 10^{-29}\,\text{J} \times \left(\frac{1\,\text{eV}}{1.6 \times 10^{-19}\,\text{J}} \right)$$

$$= -4.6 \times 10^{-10}\,\text{eV}.$$

For the parameters chosen, the interaction potential energy is comparable to the Keesom interaction discussed in Example 3.4.

3.2.4   *The interaction between two non-polar molecules*

As indicated in Fig. 3.9, a third possibility for molecule–molecule inter-actions involves two molecules (or atoms) that have neither a net charge nor permanent dipole moments. We know that such non-polar/non-polar interactions exist because inert gases like Ar condense into liquids at low temperatures. The condensation from gas to liquid results from the inter-action of transient electric dipoles that are momentarily produced by such neutral atoms or molecules. The origin of this interaction is electron delocal-ization around each atom, a quantum effect which requires wave functions that are solutions to Schrödinger's equation.

A consequence of the wave function description of electrons confined to atoms and molecules is the possibility that at any instant of time, a slight imbalance of electronic charge will be located on one side of an atomic-scale object than the other, causing a momentary (fluctuating, flickering) dipole moment that can induce a dipole moment in another nearby atom or molecule that is correlated with the first. A slight attraction will result while the momentary dipole moment exists and the electrostatic potential energy of the atomic-scale pair is lowered. Because the two dipoles are dynamically correlated, the attraction does not average to zero over time.

The likelihood that an atom or molecule produces a momentary fluctua-tion in charge density increases with size. In addition, the interaction must also be proportional to the polarizability of the second atom or molecule involved. Since polarizability is also proportional to size (see Eq. (3.24)), we might expect *a priori* that the interaction energy should be proportional to the product of the two polarizabilities.

Exact calculations of this interaction are quite involved, but a sim-ple model based on the quantum mechanical lowering of the ground state energy of two charged-coupled oscillators provides considerable physical insight into the problem [london37]. This quantum interaction is commonly referred to as the London dispersion interaction. Consider the situation shown in Fig. 3.16 which shows two non-polar molecules separated by a distance z.

Assume that a time dependent charge fluctuation in molecule 1 induces a dipole moment in molecule 2. How can we describe the effect of the charge fluctuation? Because our model system has atomic-scale dimensions, a quantum mechanical solution based on Schrödinger's equation is required. The model assumes the mass m of the charge that fluctuates is confined to a parabolic potential that can be described by an effective spring con-stant k (N/m). This gives expressions for the confining potential energy

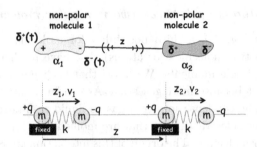

Fig. 3.16 Two non-polar molecules separated by a distance z. A momentary fluctuation in the charge distribution of molecule 1 induces a dipole moment in molecule 2. The situation is modeled by two springs with masses that are charged as shown in the bottom half of the figure. At some instant in time, the oscillating mass in molecule 1 is moving at a velocity v_1 and is separated from the fixed mass by a distance z_1 while the moveable mass in molecule 2 is moving at a velocity v_2 and is separated from the fixed mass by a distance z_2. When viewed within the context of a single molecule, q in this model is likely to be a small fraction of the single electronic charge e^-.

(Hook's law) for the charge fluctuation in each molecule

$$U_1(z) = \frac{1}{2}kz_1^2 = m\omega_o^2 z_1^2, \tag{3.50a}$$

$$U_2(z) = \frac{1}{2}kz_2^2 = m\omega_o^2 z_2^2, \tag{3.50b}$$

where ω_o (radians/s) is a characteristic oscillation frequency given by $\omega_o = \sqrt{k/m}$. If the two oscillating systems are uncoupled, Schrödinger's equation separates into two identical equations

$$\left[-\frac{h^2}{2m} \frac{\partial^2}{\partial z_1^2} \right] \Psi_1 + \frac{1}{2}m\omega_o^2 z_1^2 \Psi_1 = E_1 \Psi_1, \tag{3.51a}$$

$$\left[-\frac{h^2}{2m} \frac{\partial^2}{\partial z_2^2} \right] \Psi_2 + \frac{1}{2}m\omega_o^2 z_2^2 \Psi_2 = E_2 \Psi_2, \tag{3.51b}$$

where Ψ_1, Ψ_2 are the model-dependent wavefunctions describing the charge fluctuations and E_1 and E_2 are the energy eigenvalues of each oscillator respectively. The energy eigenvalues for these two equations are well known from any introductory quantum course[morrison10]

$$E_1 = \left(n + \frac{1}{2} \right) h\omega_o \quad n = 0, 1, 2, \ldots \tag{3.52a}$$

$$E_2 = \left(n' + \frac{1}{2} \right) h\omega_o; \quad n' = 0, 1, 2, \ldots, \tag{3.52b}$$

where n and n' are integer quantum numbers. The ground state energy is just $E_1 + E_2$ with $n = n' = 0$.

When the fluctuation involves a charge, the situation is more complicated since now the two oscillators can interact and become coupled. Using the co-ordinates defined in Fig. 3.15, the corresponding electrostatic potential energy must contain four terms and can be written as

$$U_{electr}(z) = \frac{1}{4\pi\kappa\varepsilon_o} \left[\frac{q^2}{z} - \frac{q^2}{z + z_2} - \frac{q^2}{z - z_1} + \frac{q^2}{(z - z_1) + z_2} \right]. \tag{3.53}$$

The dielectric constant κ of the surrounding medium (if any) is included to match the discussions found previously in this chapter. A consequence of this interaction term is that when $z \gg z_1$ and $z \gg z_2$, the exact electrostatic interaction in Eq. (3.53) can be approximately written as

$$U_{electr} \simeq -\frac{1}{2\pi\kappa\varepsilon_o} \frac{q^2}{z^3} z_2 z_1. \tag{3.54}$$

This gives a new expression for the potential energy in Schrödinger's Equation which is

$$U_{tot} = U_1 + U_2 + U_{elect} = \frac{1}{2}kz_1^2 + \frac{1}{2}kz_2^2 - \frac{1}{2\pi\kappa\varepsilon_o} \frac{q^2}{z^3} z_2 z_1. \tag{3.55}$$

The cross-term containing $z_2 z_1$ in Eq. (3.55) prevents a simple solution for the new energy eigenvalues of the coupled oscillator problem. However, if we rewrite the expression for U_{tot} to have a separable form, something like

$$U_{tot} = \frac{1}{2}k_s(z_1 + z_2)^2 + \frac{1}{2}k_a(z_1 - z_2)^2 \tag{3.56}$$

then progress can be made.

This can be accomplished by defining new effective spring constants k_s and k_a such that

$$\frac{1}{2}kz_1^2 + \frac{1}{2}kz_2^2 - \frac{1}{2\pi\kappa\varepsilon_o} \frac{q^2}{z^3} z_2 z_1 = \frac{1}{2}k_s \frac{(z_1 + z_2)^2}{2} + \frac{1}{2}k_a \frac{(z_1 - z_2)^2}{2}. \tag{3.57}$$

Some algebra gives

$$k_a = \left(k + \frac{1}{2\pi\kappa\varepsilon_o} \frac{q^2}{z^3} \right), \tag{3.58}$$

$$k_s = \left(k - \frac{1}{2\pi\kappa\varepsilon_o} \frac{q^2}{z^3} \right). \tag{3.59}$$

It follows that the new energy eigenvalues for the coupled system are

$$\omega_s = \sqrt{\frac{k - \frac{1}{2\pi\kappa\varepsilon_o}\frac{q^2}{z^3}}{m}},$$

(3.60a)

$$\omega_a = \sqrt{\frac{k + \frac{1}{2\pi\kappa\varepsilon_o}\frac{q^2}{z^3}}{m}}.$$

(3.60b)

The change in the quantum ground state energy will be

$$\Delta U(z) = U_{\text{charged}}(z) - U_{\text{uncharged}}(z) = \frac{1}{2}(\hbar\omega_a + \hbar\omega_s) - 2 \times \frac{1}{2}\hbar\omega_o.$$

(3.61)

Assuming that the electrostatic force $\left(\frac{1}{4\pi\kappa\varepsilon_o}\frac{q^2}{z^2}\right)$ is small compared to the restoring force of the spring $\left(\frac{1}{2}kz\right)$, we can justify a Taylor Series expansion of ω_s and ω_a. After some algebra, we find that

$$\Delta U(z) = -\frac{1}{2}\hbar\omega_o \left(\frac{1}{4\pi\kappa\varepsilon_o}\frac{q^2}{k}\right)^2 \frac{1}{z^6}.$$

(3.62)

Equation (3.62) unfortunately contains two model dependent parameters k and q which are difficult to define in a physically relevant way. These two parameters can be eliminated by realizing that a fluctuating electric field from molecule 1 causes the charge separation in molecule 2 by exerting a force which must be balanced by the "effective" spring that attempts to restore the separated charges. Essentially, the spring will "stretch" until the restoring force of the spring in molecule 2 matches the electrostatic force produced by the electric field from molecule 1 (see Fig. 3.17).

In equilibrium, this simple argument implies that

$$|q|E = kz_2.$$

(3.63)

Likewise, in equilibrium, the electric field E will induce a dipole moment in molecule 2 given by

$$p_{\text{induce},2} = |q|z_2 = \alpha E,$$

(3.64)

where α is the polarizability of the molecule in question. Evidently, by combining Eqs. (3.63) and (3.64) and using Eq. (3.23), we must have

$$\alpha = \frac{q^2}{k} = 4\pi\varepsilon_o\alpha_o.$$

(3.65)

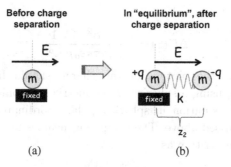

(a)   (b)

Fig. 3.17   A simple model to estimate the parameters q and k in the London dispersion force. The electric field E is produced by molecule 1 (not shown). The force of the spring in molecule 2 matches the electrostatic force produced by the electric field from molecule 1.

In this way, the two model-dependent parameters, the spring constant k and the charge fluctuation q, can be related to the polarizability of the molecule.

Example 3.6: Suppose a charge imbalance of $0.01e^-$ momentarily occurs in a He atom. Estimate the effective spring constant k required to mimic the polarizability volume of He which is measured to be about $0.2 \times 10^{-30}\,\mathrm{m}^3$.

From Eq. (3.65), we have

$$\frac{q^2}{k} = 4\pi\varepsilon_o\alpha_o$$

$$k = \frac{q^2}{4\pi\varepsilon_o\alpha_o} = \frac{[0.01 \times 1.6 \times 10^{-19}\,\mathrm{C}]^2}{4\pi\left(8.85 \times 10^{-12}\frac{\mathrm{C}^2}{\mathrm{Nm}^2}\right) \cdot \frac{1}{0.2 \times 10^{-30}\mathrm{m}^3}}$$

$$= \frac{2.6 \times 10^{-42}}{1.1 \times 10^{-10}} \cdot \frac{1}{0.2 \times 10^{-30}} = 0.12\,\frac{\mathrm{N}}{\mathrm{m}}.$$

The parameter $\hbar\omega_o$ represents a characteristic "motional" energy of the charge in a polarized atom/molecule. Lacking specific values for this quantity, we perhaps can represent it by using the characteristic ionization energy I for the atom or molecule in question. This is a well defied and easy-to-measure quantity that for chemical elements lies in the range of 5–25 eV.

With these identifications, the simple model gives rise to a result often quoted in the literature for the attractive interaction energy between two identical non-polar molecules

$$\Delta U(z) = -\frac{1}{2} \frac{\alpha^2 I}{(4\pi \kappa \varepsilon_o)^2} \frac{1}{z^6}. \tag{3.66}$$

To do better, the motion of the charge should be considered in three dimensions, using a spherically symmetric harmonic oscillator, solving Schrödinger's equation in spherical-polar coordinates rather than in 1-dimension discussed above. The energy eigenvalues for this case are well known and can be written as

$$E_n = \left(n' + \frac{3}{2}\right) \hbar \omega_o, \tag{3.67}$$

where n' is again an integer quantum number. The net result is that prefactor of $\frac{1}{2}$ in Eq. (3.66) is changed, giving

$$\Delta U(z) = -\frac{3}{2} \frac{\alpha^2 I}{(4\pi \kappa \varepsilon_o)^2} \frac{1}{z^6}. \tag{3.68}$$

For dissimilar atoms or molecules having ionization energies I_1, I_2 and polarizabilities α_1 and α_2 interacting with each other in a medium with a uniform dielectric constant κ, we finally arrive at a model for the London or dispersion interaction potential energy

$$U_{\text{London}}(z) = -\frac{3}{2} \frac{\alpha_1 \alpha_2}{(4\pi \kappa \varepsilon_o)^2} \left(\frac{I_1 I_2}{I_1 + I_2}\right) \frac{1}{z^6}, \tag{3.69}$$

where an effective ionization energy I_{eff} is often defined as

$$\frac{1}{I_{\text{eff}}} = \frac{1}{I_1} + \frac{1}{I_2} \Rightarrow I_{\text{eff}} = \left(\frac{I_1 I_2}{I_1 + I_2}\right) \tag{3.70}$$

to better approximate an "average" ionization energy when I_1 and I_2 are vastly different.

Computationally-intensive calculations of these interaction energies go well beyond the simple model discussed above and require implementations of Hartree-Fock and density functional theories which require full knowledge of the wavefunctions for the interacting atoms and/or molecules involved. A review of these efforts has recently appeared [tkatchenko10].

3.3 The van der Waals Interaction

The total attractive interaction potential energy of two freely rotating molecules separated by a distance z in a uniform dielectric medium with

Table 3.3 Dipole moments p, polarizabilities α, ionization energies I, and estimates for the various contributions to the van der Waals interaction for representative gas phase atoms and molecules. A comparison to experiment is also made where C_{Exper} is derived from the coefficients a, b found in the vdW equation of state (Eq. (2.1)). A value of $T = 300$ K was used to evaluate the Keesom contribution. The units for the various values of C are 10^{-79} Jm^6. Adapted from [butt03].

	$p(D)$	α_o (in $10^{-30} m^3$)	I(eV)	C_{Keesom}	C_{Debye}	C_{London}	C_{vdW}	C_{Exper}	Percent Diff.
He	0.00	0.20	24.6	0.0	0.0	1.2	1.2	0.9	37%
Ne	0.00	0.40	21.6	0.0	0.0	4.1	4.1	3.6	15%
Ar	0.00	1.64	15.8	0.0	0.0	51.0	5.0	45.3	13%
O_2	0.00	1.58	12.1	0.0	0.0	36.2	36.2	46.0	-21%
N_2	0.11	1.95	14.0	0.0	0.0	63.9	63.9	60.7	5%
CO	0.11	1.95	14.0	0.0	0.0	63.9	63.9	60.7	5%
HCl	0.00	2.59	12.5	0.0	0.0	100.6	100.6	103.3	-3%
HCl	1.04	2.70	12.8	9.4	5.8	112.0	127.2	156.8	-19%
CO_2	0.00	2.91	13.8	0.0	0.0	140.2	140.2	163.6	-14%
NH_3	1.46	2.30	10.2	36.4	9.8	64.7	110.9	163.7	-32%
H_2O	1.85	1.46	12.6	93.8	10.0	32.2	135.9	176.2	-23%
HBr	0.79	3.61	11.7	3.1	4.5	183.0	190.6	207.4	-8%
HI	0.45	5.40	10.4	0.3	2.2	363.9	366.4	349.2	5%
CH_3OH	1.69	3.20	10.9	65.3	18.2	133.9	217.5	651.0	-67%
$CHCl_3$	1.04	8.80	11.4	9.4	19.0	1059.4	1087.7	1632.0	-33%

dielectric constant κ is comprised of three interactions: (i) dipole–dipole, (ii) dipole–induced dipole, and (iii) induced dipole–induced dipole. Collectively these interactions are often referred to as the van der Waals interaction. From the previous sections, we have

$$U_{\mathrm{vdW}}(z) = U_{\mathrm{Keesom}} + U_{\mathrm{Debye}} + U_{\mathrm{London}}$$

$$= -\left(\frac{1}{4\pi\kappa\varepsilon_0}\right)^2 \left[\frac{1}{3}\frac{1}{k_BT}\frac{p_1^2 p_2^2}{z^6} + 2\frac{p_1^2\alpha_2}{z^6} + \frac{3}{2}\left(\frac{I_1 I_2}{I_1 + I_2}\right)\frac{\alpha_1\alpha_2}{z^6}\right]$$

$$= -\frac{C_{dW}}{z^6}. \tag{3.71}$$

Different texts may have slightly different pre-factors in front of each term, the exact value is determined by the constraints imposed at the beginning of the calculation. For instance, if the interaction between two identical atoms or molecules with dipole moment p, polarizability α, and ionization energy I, then Eq. (3.71) reduces to

$$U_{\mathrm{vdW}}(z) = -\left(\frac{1}{4\pi\kappa\varepsilon_0}\right)^2 \left[\frac{1}{3}\frac{1}{k_BT}\frac{p^4}{z^6} + 2\frac{p^2\alpha}{z^6} + \frac{3}{4}\frac{I\alpha^2}{z^6}\right]$$

$$= -\frac{C_{dW}}{z^6}. \tag{3.72}$$

It is important to realize that each term varies as z^{-6} and hence the various terms can be grouped together with one overall coefficient that depends on the properties of the interacting molecules. The London interaction is the most general of the three since it does not require either molecule to have a permanent dipole moment; it always contributes.

Table 3.3 provides a listing of the various contributions to the van der Waals interaction for a number of gas phase atoms and molecules. The table includes both calculated and experimental values and shows that the calculated values for C_{vdW} are typically within $\pm 20\%$ of experiment.

Example 3.7: Consider the interaction of two electrically neutral NH_3 molecules separated by a distance of 2 nm in vacuum.

(a) Verify the calculation for C_{vdW} for the molecule NH_3 using the values of p, α_o and I listed in Table 3.3.

(b) Evaluate the intermolecular force that two NH_3 molecules experience?

(Continued)

Example 3.7: (*Continued*)

a) We have from Eq. (3.71)

$$\langle U_{\text{vdW}}(z) \rangle = U_{\text{Keesom}} + U_{\text{Debye}} + U_{\text{London}}$$

$$= - \left(\frac{1}{4\pi\kappa\varepsilon_0} \right)^2 \left[\frac{1}{3} \frac{1}{k_B T} \frac{p_1^2 p_2^2}{z^6} + 2 \frac{p_1^2 \alpha_2}{z^6} + \frac{3}{2} \left(\frac{I_1 I_2}{I_1 + I_2} \right) \frac{\alpha_1 \alpha_2}{z^6} \right].$$

Since the two NH_3 molecules are identical, we can set $p = p_1 = p_2 = 1.46$D, $\alpha = \alpha_1 = \alpha_2 = 4\pi\varepsilon_0 \alpha_o = 4\pi\varepsilon_o (2.3 \times 10^{-30}\text{m}^3)$, and $I_1 = I_2 = I = 10.2$ eV in what follows. Since the two molecules are in vacuum, $\kappa = 1$. This gives

$$U_{\text{Keesom}} = - \left(\frac{1}{4\pi\varepsilon_0} \right)^2 \cdot \frac{1}{3} \frac{1}{k_B T} \frac{p_1^2 p_2^2}{z^6} = -\frac{C_{\text{Keesom}}}{z^6},$$

$$C_{\text{Keesom}} = \left(\frac{1}{4\pi\varepsilon_0} \right)^2 \cdot \frac{1}{3} \frac{1}{k_B T} p^4$$

$$= \frac{1}{3} \left(\frac{1}{4\pi \cdot 8.85 \times 10^{-12} \frac{\text{C}^2}{\text{Nm}^2}} \right)^2 \left(\frac{1}{1.38 \times 10^{-23} \frac{\text{J}}{\text{K}} \cdot 300\,\text{K}} \right)$$

$$\times \left(1.46\text{D} \cdot 3.33 \times 10^{-30}\text{Cm/D} \right)^4$$

$$= \frac{1}{3} \left(8.08 \times 10^{19} \frac{\text{N}^2\text{m}^4}{\text{C}^4} \right) \left(2.42 \times 10^{20} \frac{1}{\text{Nm}} \right)$$

$$\times \left(558.7 \times 10^{-120}\,\text{C}^4\text{m}^4 \right)$$

$$= 36.4 \times 10^{-79}\,\text{J}\,\text{m}^6,$$

$$U_{\text{Debye}} = - \left(\frac{1}{4\pi\kappa\varepsilon_0} \right)^2 \cdot 2 \frac{p_1^2 \alpha_2}{z^6} = -\frac{C_{\text{Debye}}}{z^6},$$

$$C_{\text{Debye}} = \left(\frac{1}{4\pi\kappa\varepsilon_0} \right)^2 \cdot 2p^2 \alpha$$

$$= \left(\frac{1}{4\pi \cdot 8.85 \times 10^{-12} \frac{\text{C}^2}{\text{Nm}^2}} \right)^2 2(1.46\text{D} \cdot 3.33 \times 10^{-30}\text{Cm/D})^2$$

(*Continued*)

Example 3.7: (*Continued*)

$$\times (4\pi \cdot 8.85 \times 10^{-12} \frac{C^2}{Nm^2} \cdot 2.3 \times 10^{-30} m^3)$$

$$= \left(8.08 \times 10^{19} \frac{N^2 m^4}{C^4} \right) (2)(23.6 \times 10^{-60} C^2 m^2)$$

$$\times \left(2.56 \times 10^{-40} C^2 \frac{m}{N} \right)$$

$$= 9.8 \times 10^{-79} J\, m^6,$$

$$U_{\text{London}} = -\left(\frac{1}{4\pi\kappa\varepsilon_0} \right)^2 \cdot \frac{3}{2} \left(\frac{I_1 I_2}{I_1 + I_2} \right) \frac{\alpha_1 \alpha_2}{z^6} = -\frac{C_{\text{London}}}{z^6},$$

$$C_{\text{London}} = \left(\frac{1}{4\pi\kappa\varepsilon_0} \right)^2 \cdot \frac{3}{2} \left(\frac{I}{2} \right) \alpha^2$$

$$= \left(\frac{1}{4\pi \cdot 8.85 \times 10^{-12} \frac{C^2}{Nm^2}} \right)^2 \frac{3}{4} \left(10.2\, eV \cdot \frac{1.602 \times 10^{-19} J}{eV} \right)$$

$$\times \left(4\pi \cdot 8.85 \times 10^{-12} \frac{C^2}{Nm^2} \cdot 2.3 \times 10^{-30} m^3 \right)^2$$

$$= \left(8.08 \times 10^{19} \frac{N^2 m^4}{C^4} \right) \frac{3}{4} (1.63 \times 10^{-18} J) \left(6.55 \times 10^{-80} C^4 \frac{m^2}{N^2} \right)$$

$$= 64.7 \times 10^{-79} J\, m^6,$$

$$C_{\text{vdW}} = C_{\text{Keesom}} + C_{\text{Debye}} + C_{\text{London}}$$

$$= (36.4 + 9.8 + 64.7) \times 10^{-79} J\, m^6$$

$$= 110.9 \times 10^{-79} J\, m^6.$$

b) The force between the two molecules is attractive and can be found by taking the negative gradient of $\langle U_{\text{vdW}}(z) \rangle$ with respect to

(*Continued*)

Example 3.7: (*Continued*)

z (see Eq. (2.11)). The force F is found to be

$$F = -\frac{\partial \langle U_{\mathrm{vdW}}(z) \rangle}{\partial z} = -\left[-C_{\mathrm{vdW}} \frac{(-6)}{z^7} \right]$$

$$= -110.9 \times 10^{-79} \mathrm{Jm}^6 \times \left(\frac{6}{(2 \times 10^{-9} \mathrm{m})^7} \right)$$

$$= -6 \left(\frac{110.9 \times 10^{-79} \mathrm{Jm}^6}{1.3 \times 10^{-61} \mathrm{m}^7} \right)$$

$$= -5.2 \times 10^{-16} \mathrm{N} = -52 \ \mathrm{fN(attractive)}.$$

Under what circumstances might such a weak force produce a measureable effect?

One possible answer is that when the vdW force is integrated over an object with dimensions of $\sim 1 \mu$ m, the vast number of atoms/molecules that must be taken into account serves as a force "multiplier" that will produce nanoNewton (nN) forces. This line of thought is the central topic of the next chapter and forms the basis for understanding the operation of an AFM.

A second answer relates to soft biological molecules like proteins which may contain thousands of polar and non-polar subunits that are placed in close proximity to each other. The charge fluctuations in such extended molecules are far greater than in individual gas-phase molecules. The net sum of 100's of tiny vdW interactions act on each segment of a protein, causing it to structurally bend and twist into a final shape that becomes biologically interesting.

3.4 Chapter Summary

In the early 1900s, understanding the origin of the non-zero interaction between electrically neutral gas-phase molecules posed a formidable challenge. An explanation for the interaction between electrically neutral atoms and molecules was ultimately reached by considering the electrical dipole moments that atoms and molecules possess. There are three origins for these dipole moments. Molecules containing atoms with different electronegativities create permanent molecular dipole moments as the atoms comprising the molecule produce small shifts in the electronic charge distribution. Secondly, molecules and atoms with no net dipole moment can develop an induced dipole moment when subjected to an external electric field. Lastly, neutral atoms and molecules with no net dipole moment can develop a fluctuating, time-dependent dipole moment that depends on the size and shape of the atom or molecule under consideration. This effect is strictly quantum mechanical in nature. The time-correlated fluctuations of these spontaneous dipole moments conspire to produce an attractive interaction between all atoms and molecules.

The derivation of the interaction potential energy between gas-phase molecules was discussed in some detail. The classical results obtained by Keesom and Debye were derived for both static dipoles and for dipoles induced by external electric fields. London's discussion of the spontaneous dipole moment in an electrically neutral object modeled as a quantum spring-mass system is also derived. While London's derivation leads to a short-range interaction that is always attractive, the interaction between permanent dipoles can be either attractive or repulsive depending on the relative dipole orientation. A thermal averaging procedure is discussed which shows how molecules with a permanent dipole moment, free to rotate, can take advantage of thermal motion to orient in such a way as to always produce a net attractive interaction. Taken together, the sum of these three possibilities — permanent dipole, induced dipole and fluctuating dipole — lead to short-range attractive interactions that all vary as $1/z^6$.

3.5   Further Reading

Chapter Three References:

[bell35]   R. P. Bell, Trans. Faraday Soc. **31**, 1557–1560 (1935).

[butt03]   H.-J. Butt, K. Graf, M. Kappl, *Physics and Chemistry of Interfaces,* Wiley-VCH, Weinheim (2003).

[debye21]   P. Debye, Z. Phyzik **22**, 302 (1921).

[griffiths13]   D.J. Griffiths, Introduction to Electrodynamics, 4th ed., Pearson, (2013).

[israelachvili98] J.N. Israelachvili, *Intermolecular and surface forces.* Academic Press, New York (1998).

[keesom21]   W.H. Keesom, Z. Physik **22**, 129 (1921).

[london37]   F. London, Trans. Farad. Soc. **33**, 8–26 (1937).

[morrison10]   John C. Morrison , *Modern Physics for Scientists and Engineers*, Academic Press (2010).
[tkatchenko10]   A. Tkatchenko, et al., MRS Bulletin **35**, 435 (2010).

Further reading:

[atkins10]   P.W. Atkins and J. de Paula, *Physical Chemistry*, 9th edition, W.H. Freeman and Co., New York NY, USA (2010).
[jeffrey 97]   G.A. Jeffrey, *An introduction to hydrogen bonding.* Oxford University Press (1997).
[kotz06]   J.C. Kotz, P. Treichel and G.C. Weaver, *Chemistry and chemical reactivity*, 6th edition, Thomson Brooks/Cole, Belmont CA, USA (2006).
[rigby86]   M. Rigby, E.B. Smith, W.A. Wakeham, and G.C. Maitland, *The forces Between Molecules,* Oxford-Clarendon Press (1986).
[stone96]   A.J. Stone, *Theory of Intermolecular Forces*, Oxford-Clarendon Press (1996).
[nelson67]   R.D. Nelson, D.R. Lide, A.A. Maryott, "Selected Values of Electric Dipole Moments for Molecules in the Gas Phase", Nat. Stand. Ref. Data Set., Natl. Bur. Stand. (NSRDS-NBS) No. 10 (1967).

3.6   Problems

1. Two identical dipoles p_1 and p_2 are separated by 1 nm in a dielectric medium with $\kappa = 5$. The dipoles are aligned in four different configurations as shown below. Complete the table below in order to specify their orientation.

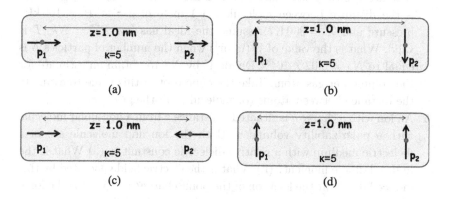

(a)   (b)

(c)   (d)

Configuration	Θ_1	Θ_2	ζ
(a)			
(b)			
(c)			
(d)			

Calculate the interaction potential energy (in eV) for each of the configurations shown above. Which of the four configurations is most stable? Which of the four configurations produces the largest force between the two dipoles?

2. Calculate the interaction potential energy between two <u>fixed</u> water molecules in vacuum whose centers are 10 nm apart, and whose dipole axes make an angle of 120° with respect to each other. Assume the water molecules lie in the same plane and they are located in free space. How does the interaction potential energy change if the two molecules can orient themselves by thermal motion? Assume $T = 300$ K.

3. A water molecule (dipole moment 1.85 D) approaches a singly charged ion. Describe the most energetically favorable orientation of the water molecule with respect to the ion. Calculate the potential energy of the interaction at a distance of 1.0 nm and compare this to the thermal energy $(3k_BT/2)$ at 20°C based on the equipartition theorem. Assume the dielectric constant of liquid water is 80. At this temperature, will the water molecule "lock-in" to a fixed position with respect to the ion?

4. The discussion in this chapter often uses distances between molecules that are typically less than 1 nm. Is this realistic? On average, what is the distance between molecules in 1 mole of gas at $P = 100$ kPa pressure and $T = 300$ K? Assume the ideal gas law $PV = Nk_BT$ is valid. What is the value of V (in m^3) when the number of particles N is equal to $N_A = 6.02 \times 10^{23}$? Divide V by N_A to obtain an estimate for the volume per gas atom. Take the cube root of this value to estimate the distance between atoms (or molecules) in the gas.

5. An ion with a charge q is situated a distance z from a non-polar molecule with a polarizability volume α. Both the ion and molecule are in a dielectric medium with a relative dielectric constant κ. (a) What is the induced dipole moment? (b) What is the electric field generated by this induced dipole at the location of the point charge? (c) What is the force of attraction between the charge q and the molecule?

6. LiCl is dissolved in water. A multitude of forces can develop as listed in the table below. Between what atomic components (Li ions, Cl ions or water molecules) do these various forces interact?

Force Component	Acts between
ion-dipole	
hydrogen bonding	
ion-ion	
ion-induced dipolar	
dipole-dipole	
dipole-induced dipolar	
induced dipole-induced dipole	

7. What do molecules of H_2, N_2, CO_2 and CH_4 have in common?

8. The interaction between two dipoles of fixed orientation that lie in the same plane is often written as

$$U_{\text{total}}(z, \theta_1, \theta_2, \zeta = 0) = \frac{1}{4\pi\kappa\varepsilon_o} \frac{1}{z^3} \left[\vec{p}_1 \cdot \vec{p}_2 - 3 \left(\vec{p}_1 \cdot \hat{r} \right) \left(\vec{p}_2 \cdot \hat{r} \right) \right]$$

$$= \frac{1}{4\pi\kappa\varepsilon_o} \frac{p_1 p_2}{z^3} \left[\cos(\theta_{12}) - 3 \cos\theta_1 \cos\theta_2 \right]$$

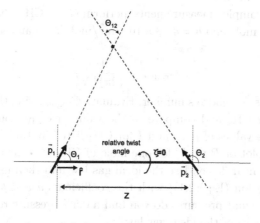

where the angles are defined in the figure. Show this is equivalent to the expression given by Eq. 3.28.

9. Calculate the potential energy of interaction between two Ar atoms separated by 0.5 nm in vacuum. The polarizability volume of Ar is about 1.6×10^{-30} m^3, the ionization energy of Ar is about 15.7 eV per atom.

10. Two methane molecules interact via a 12-6 Lennard-Jones potential characterized by the parameters $\varepsilon = 2.1 \times 10^{-21}$ J and $\sigma = 0.42$ nm. Derive an equation for the force on a methane molecule as a function of its distance from another methane molecule. Make a plot of the force vs. separation. Make sure the force has units of nN and the separation has units of nm. What is the separation between the two molecules when the net force acting on the two methane molecules becomes zero?

11. A water molecule has a polarizability volume of 1.5×10^{-30} m^3 and a permanent dipole moment equal to 1.85 D. At what separation distance will a Na ion induce a dipole moment in a water molecule equal to its permanent dipole moment? Assume the dielectric constant of liquid water is 80, a constant independent of the molecule-molecule separation.

12. The coefficients appearing in the vdW equation of state for a gas (see Eq. (2.1)) can be used to define an experimental value for C_{vdW}. The coefficients a, b are determined by fitting P–V data for a gas. Ideally, you might expect $PV = nRT$. But careful experiment shows that P is reduced from the value predicted by the ideal gas law. By fitting the measured P–V data, accurate values for the vdW coefficients a and b can be obtained.

    As an example, measurements performed on CH$_4$ show that $a = 0.225$ Nm^4/mol^2 and $b = 4.28 \times 10^{-5}$m^3/mol. You can use these values to define

    $$C_{\mathrm{Exper}} = \frac{9ab}{(2\pi)^2 N_A^3},$$

    where N_A is Avagadro's number. Evaluate C_{Exper} using the coefficients a and b for CH$_4$ and compare to the value of C_{vdW} found using the appropriate values of p, α, and I for CH$_4$ listed in Table 3.3.

    Make a plot of P vs. V for one mole of CH$_4$ at some constant temperature T near 300 K. Use the ideal gas law and then use the van der Waals equation (Eq. (2.1)) with the coefficients a and b given above. Roughly at what pressures do you find a 0.5% pressure reduction from the predictions of the ideal gas law?

13. From P–V measurements, it is known that water vapor does not strictly obey the ideal gas law. This implies that water molecules attract each other when in the gas phase. Roughly, what percentage of this attraction is due to the London dispersion force?

14. The heat of vaporization (or the enthalpy of vaporization) measures the amount of energy required to turn a fixed amount of liquid at

its normal boiling temperature into a gas at the same temperature. Ultimately, the value of the heat of vaporization must be related to the strength of interaction between nearest neighbor atoms. Use the value of C_{vdW} from Table 3.3 to roughly estimate the energy required to convert one mole of liquid neon into gas at its boiling temperature of 27 K. Assume the inter-atomic Ne-Ne distance is about 0.3 nm. At first, assume a Ne atom only interacts with another Ne atom. How does your estimate compare to the measured value of 1.7 kJ/mole? In order to better match your estimate to measurements, roughly how many Ne atoms must interact in the liquid state?

15. Noble gas atoms are inert. They are odorless, colorless, non-flammable, and have low chemical reactivity. If atoms cannot interact, are they electronegative? The inert behavior demonstrated by all noble gasses is a common characteristic suggesting that any noble gas atom-atom interaction might be dominated by the London dispersion force. If separate containers of the noble gasses were filled to atmospheric pressure at 25°C, which noble gas would you predict might be the most strongly interacting? For reference, the boiling points and atomic radii of the noble gases are listed in the table below. Make a plot of boiling point vs. atomic size. Is there a correlation?

Noble Gas	Chemical Symbol	Boiling Point	Atomic radius
helium	He	4.2 K	31 pm
neon	Ne	27.3 K	38 pm
argon	Ar	87.4 K	71 pm
krypton	Kr	121.5 K	88 pm
xenon	Xe	166.6 K	108 pm
radon	Rn	211.5 K	120 pm

Chapter 4

Van der Waals Interactions
between Macroscopic Objects

The van der Waals (vdWs) interactions between individual atoms discussed in the previous chapter have three distinct origins: dipole–dipole, dipole–induced dipole, and induced dipole–induced dipole. Two important features come from a careful study of these interactions: their short range and the small net attractive force that can develop between atoms. In 1937, Hamaker developed a theory for vdW interactions between macroscopic bodies and showed that if pair-wise additivity of inter-molecular vdW forces is assumed, the vdW interaction can lead to long-range forces with sufficient strength to make them directly observable in a laboratory setting [hamaker37]. In 1959, a more fundamental theory was developed by Dzyaloshinskii, Lifshitz and Pitaevskii, which circumvented the pair-wise additive assumption [dzyaloshinskii61].

In the following chapter we review these various theories and use them to derive equations describing the interaction of a tip with a flat substrate, a geometry of considerable interest to AFM applications.

To use an AFM intelligently, you must be aware of the models that describe the force between a tip and substrate and you must understand what features of these models you can control experimentally. The predictions of these models are routinely used to quantitatively understand AFM data. In general terms, these topics are usually of most interest to students of mechanical engineering and physics.

4.1   Integrating the van der Waals Interaction

In Chapter 3, we discussed the vdW interaction potential energy between two atoms that are separated by a distance z in a medium held at temperature T having a dielectric constant κ. If the atoms have dipole moments p_1, p_2, polarizabilities α_1, α_2 and ionization energies I_1 and I_2, the interaction potential energy can be written as

$$U_{\text{vdW}}(z) = U_{\text{Keesom}} + U_{\text{Debye}} + U_{\text{London}}$$

$$= -\left(\frac{1}{4\pi\kappa\varepsilon_0}\right)^2 \left[\frac{1}{3}\frac{1}{k_B T}\frac{p_1^2 p_2^2}{z^6} + 2\frac{p_1^2 \alpha_2}{z^6} + \frac{3}{2}\left(\frac{I_1 I_2}{I_1 + I_2}\right)\frac{\alpha_1 \alpha_2}{z^6}\right]$$

$$= -\frac{C_{\text{vdW}}^{(1,2)}}{z^6}. \tag{4.1}$$

The first step to extend this result is to consider how an atom located a distance d from a homogeneous flat plate interacts with **all** the atoms in that plate. The situation is sketched in Fig. 4.1. To calculate the net interaction, we must sum the atom-atom interaction in Eq. (4.1) over all the atoms in the half plane. In principle, this situation is different from the discussion in Chapter 3, where only the interactions between individual atoms were discussed.

By broadening our scope to include atoms in the condensed state, new effects can arise due to interactions with nearest neighbor atoms. For amorphous solids, you might suspect these interactions tend to average out, but this need not be the case for crystalline solids. When we casually write about "summing over all the atoms in the solid" we really mean to say "summing over all the unit cells in the crystalline solid". Depending on the

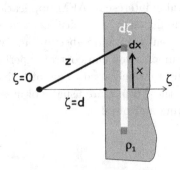

Fig. 4.1   The geometry required to calculate the interaction potential energy of an atom located a distance d from a large flat half plane which we call the substrate.

symmetry and atomic structure of the unit cell, new issues related to the condensed state of matter can become important that are not considered here. These issues will be considered in Sec. 4.4 below.

With this caveat, all the atoms located in a ring of radius x, thickness dx, and width $d\zeta$ are an equal distance z from the external atom located at $\zeta = 0$. All contribute equally to the interaction potential. How many atoms N are in the ring? The number density ρ_1, of atoms (m^{-3}) in the plate is given by

$$\rho_1 = \frac{N_A}{m} \rho_{\text{mass}}, \tag{4.2}$$

where N_A is Avagadro's number (6.02×10^{23}), m is the mass of each atom (kg) comprising the substrate (assumed all identical) and ρ_{mass} is the mass density of the substrate (kg/m^3). Typically, ρ_1 lies in the range between $\sim 2 - 15 \times 10^{28}$ m^3 for the elements in the periodic table.

The volume dV of the ring is $2\pi x dx\, d\zeta$. This leads to a contribution to the interaction potential given by

$$dU = -N \frac{C_{\text{vdW}}^{(1,2)}}{\left(\sqrt{\zeta^2 + x^2}\right)^6} = -\frac{2\pi C_{\text{vdW}}^{(1,2)} \rho_1 x dx\, d\zeta}{(\zeta^2 + x^2)}. \tag{4.3}$$

The total interaction will be given by integrating over all the atoms comprising the entire plate.

$$U_{\text{vdW}}(d) = \int dU = -2\pi \rho_1 C_{\text{vdW}}^{(1,2)} \int_{\zeta=d}^{\zeta=\infty} d\zeta \left(\int_{x=0}^{x=\infty} \frac{x\, dx}{(\zeta^2 + x^2)^3} \right)$$

$$= -2\pi \rho_1 C_{\text{vdW}}^{(1,2)} \int_{\zeta=d}^{\zeta=\infty} \left(\frac{1}{4} \times \frac{1}{\zeta^4} \right) d\zeta = -\frac{2\pi \rho_1 \dot{C}_{\text{vdW}}^{(1,2)}}{4} \left(-\frac{1}{3\zeta^3} \right) \Big|_{\zeta=d}^{\zeta=\infty}$$

$$U_{\text{vdW}}(d) = \frac{-\pi \rho_1 C_{\text{vdW}}^{(1,2)}}{6d^3}. \tag{4.4}$$

The next step is to approximate a sharp tip as a sphere of radius R that is positioned a distance d in front of the flat plane. We must now integrate the interaction found in Eq. (4.4) over the geometry of the sphere. As shown in Fig. 4.2(a), this is accomplished by defining a variable of integration ζ such that $0 \leq \zeta \leq 2R$. According to Eq. (4.4), an atom located a distance $d+\zeta$ from the flat plane will experience an interaction potential dU

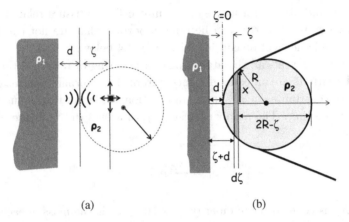

Fig. 4.2 In (a), a single atom at a distance $d+\zeta$ above an infinite flat block of material. The block has an atom number density of ρ_1. The individual atom interacts with the block by van der Waals forces. In (b), a sphere–plane interaction develops by integrating the van der Waals force over the geometry of the sphere. A tip used in AFM is often approximated by a sphere with a well-defined radius R. To safely make this assumption, the separation distance d must be much greater than any atomic details on the sphere or block.

given by

$$U_{\mathrm{vdW}}(d+\zeta)|_{\text{one atom}} = \frac{-\pi\rho_1 C_{\mathrm{vdW}}^{(1,2)}}{6(d+\zeta)^3}. \tag{4.5}$$

Figure 4.2(b) defines the geometry required to integrate over the sphere of radius R. The shaded slab of infinitesimal thickness $d\zeta$ has a volume dV_{slab} given by

$$dV_{\text{slab}} = \pi x^2 d\zeta = \pi[R^2 - (R-\zeta)^2]d\zeta = \pi(2R-\zeta)\zeta d\zeta. \tag{4.6}$$

The number of atoms in the shaded slab, denoted by N_2, is given by

$$N_2 = \rho_2 dV_{\text{slab}}, \tag{4.7}$$

where ρ_2 is the number density of atoms in the tip. All the atoms in the shaded slab are an equal distance $(d+\zeta)$ from the substrate, so the net interaction potential can be obtained by integrating the slab over the tip having a shape that we approximate by the sphere of radius R.

$$\begin{aligned} U(d) &= -\frac{\pi\rho_1 C_{\mathrm{vdW}}^{(1,2)}}{6} \int_{\zeta=0}^{\zeta=2R} \frac{\rho_2 dV_{\text{slab}}}{(d+\zeta)^3} \\ &= -\frac{\pi\rho_1 C_{\mathrm{vdW}}^{(1,2)}}{6} \cdot (\pi\rho_2) \int_{\zeta=0}^{\zeta=2R} \frac{(2R-\zeta)\zeta d\zeta}{(d+\zeta)^3}. \end{aligned} \tag{4.8}$$

With some effort, the integral in Eq. 4.8 can be solved analytically to give

$$U(d) = -\frac{\pi^2 \rho_1 \rho_2 C_{vdW}^{(1,2)}}{6}$$
$$\times \left[\frac{R}{d} + \frac{3}{2} + \ln\left(\frac{d}{d+2R}\right) - \left(\frac{2R(d+4R)+d(3d+8R)}{2(d+2R)^2}\right) \right]$$

If $R \ll d$, the term

$$\frac{2R(d+4R)+d(3d+8R)}{2(d+2R)^2} \to \frac{3}{2}$$

and

$$U(R \ll d) = -\frac{\pi^2 \rho_1 \rho_2 C_{vdW}^{(1,2)}}{6} \left[\frac{R}{d} + \ln\left(\frac{d}{d+2R}\right) \right].$$

If $R \gg d$, a situation more relevant to AFM, we have

$$U_{\text{plane-sphere}}(R \gg d) = -\frac{\pi^2 \rho_1 \rho_2 C_{vdW}^{(1,2)}}{6} \left[\begin{array}{l} \frac{R}{d} + \frac{1}{2} + \ln\left(\frac{d}{d+2R}\right) \\ -\frac{1}{4}\frac{d}{R} + \frac{1}{8}\left(\frac{d}{R}\right)^2 + \cdots \end{array} \right]$$

$$= -\frac{H_{1,2}}{6} \left[\frac{R}{d} + \frac{1}{2} - \ln\left(1+\frac{2R}{d}\right) \right.$$
$$\left. -\frac{1}{4}\frac{d}{R} + \frac{1}{8}\left(\frac{d}{R}\right)^2 + \cdots \right] \tag{4.9}$$

The quantity $H_{1,2} = \pi^2 \rho_1 \rho_2 C_{vdW}^{(1,2)}$ (often represented as H or A or A_H in the literature) is known as Hamaker's constant and is a number that quantifies the strength of the interaction potential between a sphere of radius R and a large flat substrate. Tables for different materials appear throughout the literature [visser72], [bargeman72], [croucher77], [croucher81], [bergstrom97]. Values of $H_{1,2}$ for materials commonly used in AFM experiments are also compiled in focused reviews [capella99], [leite12].

The final result given in Eq. (4.9) is remarkable because the short range $1/z^6$ behavior for the vdW interaction in Eq. (4.1) integrates out and produces a much longer range d^{-1} result. The force between the sphere and flat substrate is given by taking the negative derivative of Eq. (4.9) with respect to d. The leading R/d term dominates when $R/d \gg 1$, giving

$$F_{\text{plane-sphere}} = -\frac{\partial U_{\text{plane-sphere}}(d)}{\partial d} = -\frac{H_{1,2}R}{6d^2} \qquad (4.10)$$

The negative sign indicates that the force is attractive. In what follows, we will often replace $H_{1,2}$ by simply H and assume that the appropriate value of H for materials 1 and 2 is selected.

Example 4.1: Estimate the order of magnitude of the Hamaker constant $H_{1,2}$ for a typical substrate/tip combination and compare to $k_B T$, a typical thermal energy. Use the estimate of $H_{1,2}$ to calculate the force that a sphere of radius $R = 6\,\text{nm}$ will experience when placed a distance $d = 1\,\text{nm}$ above a flat substrate.

Solution: Let's assume the number density of atoms in the tip and substrate are the same ($\rho = \rho_1 = \rho_2$) and that ρ spans a typical range of values given by $(2 - 15) \times 10^{28}\,\text{m}^{-3}$. A table of exact values for the number density of the chemical elements can be found in the literature [kittel86]. From the results in Table 3.3, the vdW constant C_{vdW} typically lies in range between $10 - 100 \times 10^{-79}\,\text{Jm}^6$. These assumptions give a range of values for $H_{1,2}$

$$H_{1,2} = \pi^2 \rho_1 \rho_2 C_{\text{vdW}} = \pi^2 \left(\left\{ \begin{matrix} 15 \\ 2 \end{matrix} \right\} \times 10^{28}\,\text{m}^{-3} \right)^2 \left(\left\{ \begin{matrix} 100 \\ 10 \end{matrix} \right\} \times 10^{-79}\,\text{Jm}^6 \right)$$

$$\simeq \left(\left\{ \begin{matrix} 2 \times 10^{-18} \\ 4 \times 10^{-21} \end{matrix} \right\}\,\text{J} \right) \cdot \left(\frac{1\,\text{eV}}{1.6 \times 10^{-19}\,\text{J}} \right) = \left\{ \begin{matrix} 14 \\ 0.025 \end{matrix} \right\}\,\text{eV}$$

The interaction potential energy is given by Eq. (4.9) and for $R = 6\,\text{nm}$ and $d = 1\,\text{nm}$, we have $U = -H_{12} = -\left\{ \begin{matrix} 14 \\ 0.025 \end{matrix} \right\}\,\text{eV}$. Thermal energies at $T = 300\,\text{K}$ are given by

$$k_B T \simeq 1.38 \times 10^{-23}\,\text{J/K} \cdot 300\,\text{K} \simeq 4 \times 10^{-21}\,\text{J} = 0.026\,\text{eV}.$$

We thus see that the interaction between two macroscopic bodies can be comparable to or larger than thermal energies. This situation should be contrasted to the case of the interaction of two individual atoms discussed in Chapter 3.

(Continued)

Example 4.1: (*Continued*)

The force that a spherical tip of radius $R = 6\,\text{nm}$ will experience when placed a distance $d = 1\,\text{nm}$ above a flat substrate is attractive and can be estimated from Eq. (4.10)

$$|F_{\text{vdW}}| = \frac{H_{1,2}R}{6d^2} = \frac{\left(\left\{\begin{array}{c} 2 \times 10^{-18} \\ 4 \times 10^{-21} \end{array}\right\}\,\text{J}\right)(6 \times 10^{-9}\text{m})}{6(1 \times 10^{-9}\text{m})^2}.$$

$$= \left\{\begin{array}{c} 2 \times 10^{-9} \\ 4 \times 10^{-12} \end{array}\right\} N = \left\{\begin{array}{c} 2\,\text{nN} \\ 4\,\text{pN} \end{array}\right\}.$$

The force ranges from pN to nN depending on the details of the tip/substrate composition. Forces of this magnitude can be measured by an AFM.

It should be clear that the specific force law derived from van der Waals interactions (see Eq. (4.9)) depends critically on the geometry of the interacting objects. Israelachvili's book presents a summary of the force laws for a variety of different geometries [israelachvili98].

There are a number of approximations that have been glossed over when deriving the Hamaker result stated in Eq. (4.9). One assumption replaces the summation over all interacting pairs of atoms by integration over a well-defined geometry. For this assumption to be strictly valid, the separation between the interacting bodies must be large enough so that they may be treated as continuous geometric objects and not a collection of discrete particles. Furthermore, atomic homogeneity is assumed, an assumption that is certainly not true near surfaces where oxidation and contamination layers can build up. Therefore, the Hamaker result becomes suspect at separations less than a few atomic diameters, where the atomic nature of matter becomes an important factor.

Another drawback of the Hamaker approach is the neglect of the propagation time for the electric field produced by a fluctuating dipole. Properly accounting for this propagation time (known as retardation effects) becomes important when the time delay is comparable to the fluctuation time of the dipole itself. In general, retardation effects become more important when surfaces interact through an intermediate liquid in which the speed of light is reduced or when the two materials are separated by a large distance.

Lastly, inherent in the Hamaker approach is the assumption of **pairwise-additivity**. This assumption is widely invoked because of its inherent simplicity. However, this approach ignores the influence of neighboring atoms on the interaction between any pair of atoms. For instance, pairwise additivity assumes the polarizability of an atom is not affected by other nearby atoms. However, any electric field produced by atom 1 interacts directly with atom 2, but also indirectly with atom 2 through interactions with all neighboring atoms. While pairwise-additivity is probably a good approximation when discussing gases, when dealing with condensed matter, this assumption can lead to inaccuracies. Thus you might expect that a simple sum of all pair-wise interactions will predict a force greater than the actual force between two objects separated by a small distance.

In light of the above discussion, it is perhaps not surprising that the values of the Hamaker constant measured by a wide variety of techniques are somewhat inconsistent.

It is also worth commenting that the van der Waals interaction is routinely invoked in the literature to explain a wide variety of physical phenomena often without any proof. A convincing proof that van der Waals forces are important in a given situation ultimately relies on a careful measurement of force vs. separation, a measurement which is often difficult to perform. Thus, for example, it is questionable to claim that sub-micron particulates in solution *always* flocculate and precipitate due to vdW forces. To be precise, such a claim is not valid unless a direct measure of force vs. distance is somehow made between the sub-micron particles.

4.2   Surface Energy and Adhesion and Its Relation to Hamaker Constants

Surfaces and interfaces are generally defined as the boundary region between two adjacent phases of matter. When discussing gas/solid, vapor/liquid or liquid/solid interactions, it is customary to designate the interacting boundaries as *surfaces*. When solid/solid or liquid/liquid boundaries are under discussion, it is customary to call the interacting boundaries *interfaces*. When a tip comes into close proximity to a substrate, any resulting physical and/or chemical processes that might occur depends on the strength and range of interaction between relevant atoms in the tip and substrate. Thus when the tip is separated from a substrate by a thin layer of air, we focus on *surface* interactions, while if the tip comes into physical contact with the substrate, the *interface* that forms becomes the topic of interest.

Atoms at an interface are in a state of higher energy than those in the bulk due to a lack of nearest neighbor interactions. For this reason, it is well known from surface science that atoms at a clean surface tend to be under compression or tension that is sufficiently strong and complex to produce a staggering variety of different atomic corrugations. The resulting interaction is often dominated by surface defects present either on the tip and/or on the substrate. To describe the physics behind these interactions, the concept of a surface energy was developed by Thomas Young in 1805, at a time coincident with the emerging view that atoms form the basis of all matter. Over the years, the concept of surface energy has been reinterpreted to include an atomistic point of view. Since the concept has survived, we must make an attempt to understand the physics behind it and learn how to correlate it with the vdW interactions discussed above.

4.2.1   Creating a flat surface

An important case to consider is the vdW interaction of two parallel surfaces separated by a distance d from each other. To discuss this case, we first consider the energetics required to produce such a configuration.

The differential work dW required to create a differential surface area dA must be proportional to the number of atoms in dA and therefore dW must be proportional to dA. To make the proportionality an equation, we define a proportionality constant γ such that

$$dW = \gamma dA, \qquad (4.11)$$

where γ is a scalar quantity with units of energy/unit area (J/m^2). Thus γ is known as the specific surface energy of a particular material. It is sometimes useful to think of γ as a restoring force per unit length that resists an increase in area. In the case of liquids, γ is numerically equal to the surface tension which is a vector with units of force/unit length, (N/m). It is thus common to say that surface tension in liquids acts to decrease the surface energy of the system. This simple principle explains well-known geometric results like the spherical shape of a suspended liquid droplet or the characteristic crescent-like shape (meniscus) of a liquid-vapor interface in small capillary tubes.

From a microscopic point of view, γ must roughly equal the interaction energy of an atom with its nearby neighbors divided by the area per atom.

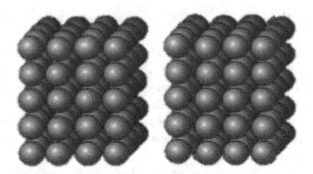

Fig. 4.3   An atomic-scale representation of two perfect surfaces of equal area dA created by cleaving a solid.

If the interaction energy is of order $k_B T$ (weakly interacting atoms) and the "size" of an atom is of order 0.5 nm, at room temperature the magnitude of γ should be about

$$\gamma \approx \frac{1.38 \times 10^{-23} \, \text{J/K} \times 300 \, \text{K}}{(0.5 \times 10^{-9})^2} \approx 0.02 \, \text{J/m}^2. \qquad (4.12)$$

Since γ determines the shape of a liquid drop, there are many ways that have been devised to measure γ for liquids. How to measure γ for a solid is more problematic. In principle, the work required to produce a solid–solid interface could be measured when a solid is cleaved, creating two new surfaces each with area dA as shown schematically in Fig. 4.3. However any work that is required to cleave a solid will certainly be in excess of the minimal work required due to a multitude of uncontrollable events that can occur during such a violent process.

In fact, the process of creating a solid–solid interface is complicated as schematically indicated in Fig. 4.4 where a perfect interface, once formed, is usually followed by a spontaneous reorganization of the surface atoms to positions of lower energy.

The energy required to perfectly cleave a crystalline surface should be proportional to the number of bonds that are broken as the new surface is created. This quantity can be estimated if the arrangement of atoms in the bulk is known with some certainty. For example, the number of bonds broken could be estimated by simply counting the number of nearest interacting neighbors before cleaving.

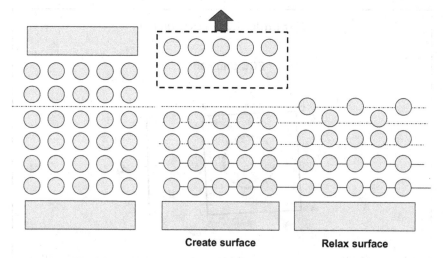

Create surface   **Relax surface**

Fig. 4.4 When a new solid interface is created, the re-arrangement of atoms at the newly created surface is driven by an energy minimization process.

If the surface contains a number density ρ_j of atoms of type j which forms n_j bonds of energy ε_j with its nearest neighbors, then the surface energy must be fundamentally given by an equation of the form

$$\gamma = \frac{1}{2}\sum_j \rho_j n_j \varepsilon_j, \tag{4.13}$$

where the summation index j runs over all possible atom types found on the cleaved surface. The factor of $1/2$ accounts for the fact that two surfaces of equal surface energy are formed.

Example 4.2:   Estimate the surface energy of a crystalline solid that condenses in the face-centered cubic (FCC) structure. Assume a (100) surface is formed and that all atoms in the unit cell are identical.

The unit cell of an FCC structure is shown in the diagram where the rectangle highlights the (100) surface. The lattice constant a is shown.

(Continued)

Example 4.2: (*Continued*)

FCC unit cell

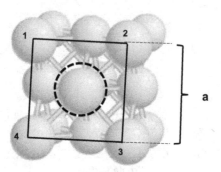

When a (100) surface is formed, some atoms in this unit cell will be removed to form an opposing plane. Assume on average, the atom indicated by the dotted circle is removed from each unit cell. This atom has four nearest neighbors as shown with bond energies ε.

Since all atoms in the unit cell are identical, the equation for γ can be simplified

$$\gamma = \frac{1}{2} \sum_{j=1} \rho_j n_j \varepsilon_j = \frac{1}{2}\rho n \varepsilon$$

$$\rho = \frac{2\,\text{atoms}}{a^2}$$

$$n = 4$$

$$\gamma = \frac{4\varepsilon}{a^2}$$

Thus, in principle, if the bond energy ε and the surface structure are known, the surface energy γ can be estimated.

A table listing various values of γ for a few common materials is given in Table 4.1. General trends for surface energies indicate relatively high values ($\sim 1000\,\text{mJ/m}^2$) for metals and relatively low values ($\sim 30\,\text{mJ/m}^2$) for liquids.

A plot of surface energies of the solid chemical elements is provided in Fig. 4.5 and shows the trends in surface energy across the periodic table.

Table 4.1   Typical values of surface energies for a number of common solids and liquids. The materials are listed roughly in order of decreasing values for γ. An exhaustive list for other materials can be found in the literature [jasper72], [bauer96].

SOLIDS	Surface energy (mJ/m^2)	LIQUIDS	Surface energy (mJ/m^2)
mica (in vacuum)	4500		
mica (in air)	350		
copper	1100		
gold	1000		
aluminum	840	mercury	487
glass	40–100		
nylon	46	water	73
polystyrene	41	benzene	29
acrylic	38	acetone	25
polyethylene	31	methanol	23
teflon	18	ethanol	22

It is evident that the transition metals have the highest surface energies among the elements.

In general, solid materials with high surface energy tend to exhibit strong cohesion, have high melting temperatures, and tend to rapidly adsorb contaminants. Reliable values for the surface energy of a specific solid are often difficult to measure since surface oxides, adsorbates, surface roughness, and surface treatment can significantly affect experimental values.

It is important to appreciate that there can be many contributions to the surface energy of a particular material which is represented by a single number found in the literature. *A priori*, two surfaces when brought into contact to form an interface, can interact and stick together for a number of reasons which include mechanical, chemical, dispersive (London), electrostatic and diffusive effects.

Mechanical interactions result if one of the surfaces might have an adhesive-like viscosity capable of filling pores and voids to produce mechanical interlocking.

Chemical interactions between atoms can form chemical bonds (either hydrogen bonds or covalent/ionic bonds) across an interface, binding the two surfaces together.

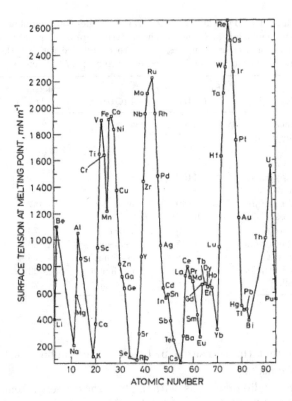

Fig. 4.5   The surface energy of solid chemical elements at their melting point [keene93].

Dispersive adhesion takes place when atoms in each surface attract one another by van der Waals forces.

In surface science, the term "adhesion" almost always refers to dispersive adhesion alone, a process which is also known as physisorption.

Electrostatic adhesion results in an exchange of electrons across an interface, producing an interfacial structure compressed by electrical interactions similar to those found in a parallel plate capacitor.

Finally, for the case of diffusive adhesion, if one or both of the surfaces is comprised of polymer-like molecules that become entangled one with the other, an interfacial bond can be formed by diffusion.

As an example, the surface energy of H_2O ($\gamma = 73 \, \text{mJ/m}^2$) is thought to be comprised primarily of two contributions:

$$\gamma_{H_2O} = \gamma_{\text{dispersion}} + \gamma_{H-\text{bond}}$$
$$= 22 \, \text{mJ/m}^2 + 50.8 \, \text{mJ/m}^2 = 72.8 \, \text{mJ/m}^2 \qquad (4.14)$$

It follows that the adhesion of two unlike materials like a tip and substrate is largely determined by the force of attraction between the atoms that comprise the different materials. The case when one of the materials happens to be a liquid is easiest to discuss first.

4.2.2   *The liquid-solid interface*

The surface energetics of the liquid-solid interface is not of much relevance in AFM, so they will not be discussed in any great detail here. Suffice it to say that if a drop of liquid is placed on two solid substrates, one with a high surface energy and one with a low surface energy, the atoms of the liquid will experience a different force of attraction between the two substrates. If the substrate has a high surface energy, the atoms in the liquid will be strongly attracted to the atoms of the substrate and the drop will spread across the solid surface. If the substrate has a low surface energy, the atoms of the liquid will be attracted to each other more than to the atoms of the solid surface and the liquid drop will not spread. Observing the degree to which a drop of high purity water spreads across a metal surface thus provides a very simple qualitative test to determine the cleanliness of that surface. If the metal surface is highly contaminated with hydrocarbons, a water drop will remain a drop and will not spread to cover the surface.

4.2.3   *The solid–solid interface*

The AFM tip and the substrate are typically made from dissimilar solid materials. It is therefore worthwhile to extend the discussion above to the case of solid–solid interfaces which produce two different solid surfaces when separated. An important question to answer is how can we describe the work (energy) required to separate a tip from a substrate in terms of the surface energies of the tip and substrate alone? The situation is shown in Fig. 4.6 in which two dissimilar materials, 1 and 2, form an interface.

The difficulty posed here is that solids do not deform easily, so the effect of surface energies is difficult to measure. This situation should be contrasted to liquids where a small change in surface energy produces measureable shape changes. Historically, that is why surface energies for solids are usually obtained from the molten phase (see Fig. 4.5).

There can be errors if the surface energies inferred at high temperatures are assumed to be temperature independent. As temperature increases, the atoms in a solid vibrate more. Because of the increased vibration, the

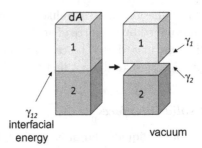

Fig. 4.6   Two dissimilar materials are joined to make a solid–solid interface. After separation, two dissimilar surfaces are formed.

average distance between the atoms increases and the average force holding them together decreases. If the attractive forces decrease, the amount of work required to separate the atoms goes down, hence so does the surface energy. Thus results for γ obtained near the melting point of a solid may be inaccurate when they are used to estimate the surface energy of a solid near room temperature.

Following Fig. 4.6, the work required to separate two dissimilar materials and create two surfaces of area dA can in principle be related to the difference between the surface and interfacial energies according to

$$dW_{12} = (\gamma_1 + \gamma_2 - \gamma_{12})dA. \qquad (4.15)$$

Often the symbol $\Delta\gamma$ is used to designate the quantity $\gamma_1 + \gamma_2 - \gamma_{12}$. When the two solids are different, dW_{12} is called the work of adhesion. While γ_1 and γ_2 might be found in tables, it is more difficult to know how to determine γ_{12} for a specific interface. In principle, this important quantity can be found from the Hamaker theory discussed in Sec. 4.2 above.

If the surface-surface interaction is due solely to the London dispersion interaction (i.e., no chemical bonds are formed at the interface), one can perform a Hamaker calculation for two dissimilar materials separated by a distance d that have flat surfaces parallel to each other. The geometry is sketched in Fig. 4.7.

Assume that material 1 is infinitely large and has an atom number density ρ_1 and material 2 has an area A and is made from a material with an atom number density ρ_2. Recall that the interaction of an atom located a distance $d + \zeta$ away from a flat infinite plane is given by Eq. (4.5) as

$$U_{\mathrm{vdW}}(d) = \frac{-\pi\rho_1 C_{\mathrm{vdW}}^{(12)}}{6(d + \zeta)^3} \qquad (4.16)$$

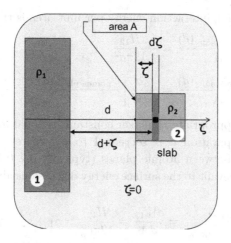

Fig. 4.7   The interaction of two dissimilar materials 1 and 2 when separated by a distance d. Material 1 (with atom density ρ_1) is taken to be an infinitely wide flat plane. An integration of the vdW interaction over the geometry of material 2 (with atom density ρ_2) is required.

where $C_{\text{vdW}}^{(12)}$ is the appropriate van der Waals constant that describes the interaction between atoms of type 1 and atoms of type 2. This interaction characterizes all the atoms in the shaded slab of thickness $d\zeta$ in Fig. 4.7, which is at a constant distance of $d + \zeta$ from the infinitely large substrate.

Since the number of atoms in the slab is $\rho_2 dV_{\text{slab}} = \rho_2 A d\zeta$, we have

$$U_{\text{plane}-\text{plane}}(d) = -\frac{\pi \rho_1 C_{\text{vdW}}^{(12)}}{6} \int_{\zeta=0}^{\zeta=\infty} \frac{\rho_2 dV_{\text{slab}}}{(d+\zeta)^3}$$

$$= -\frac{\pi \rho_1 \rho_2 C_{\text{vdW}}^{(12)}}{6} \int_{\zeta=0}^{\zeta=\infty} \frac{A d\zeta}{(d+\zeta)^3}$$

$$= -\frac{\pi \rho_1 \rho_2 C_{\text{vdW}}^{(12)}}{6} A \left(-\frac{1}{2(d+\zeta)^2} \right) \Bigg|_{\zeta=0}^{\zeta=\infty}$$

$$= -\frac{\pi \rho_1 \rho_2 A C_{\text{vdW}}^{(12)}}{6} \left(\frac{1}{2d^2} \right) \tag{4.17}$$

The plane–plane interaction energy per unit area is then evidently

$$\frac{U_{\text{plane}-\text{plane}}(d)}{A} = -\frac{H_{12}}{12\pi d^2} \tag{4.18a}$$

$$\frac{F_{\text{plane}-\text{plane}}(d)}{A} = -\frac{\partial}{\partial d}\left[\frac{U_{\text{plane}-\text{plane}}(d)}{A}\right] = -\frac{H_{12}}{6\pi d^3} \tag{4.18b}$$

where H_{12}, the appropriate Hamaker constant, is equal to $\pi^2 \rho_1 \rho_2 C_{\text{vdW}}^{(12)}$.

When the separation d is set equal to a_o where a_o is the equilibrium separation between atomic planes (typically 0.2–0.3 nm), we should recover results relevant to the surface energy discussions above. This implies that

$$\left.\frac{U_{\text{plane}-\text{plane}}}{A}\right|_{d=a_o} = \frac{dW_{12}}{dA} = \frac{H_{12}}{12\pi a_o^2} \simeq \gamma_1 + \gamma_2 - \gamma_{12}. \tag{4.19}$$

If material 2 is the same as material 1, then $\gamma_1 = \gamma_2$ and $\gamma_{12} = 0$, giving

$$\left.\frac{U_{\text{plane}-\text{plane}}}{A}\right|_{d=a_o} = \frac{dW_{11}}{dA} = \frac{H_{11}}{12\pi a_o^2} \simeq 2\gamma_1. \tag{4.20}$$

This provides an expression that illustrates the fundamental origin of the surface energy

$$\gamma_1 = \frac{H_{11}}{24\pi a_o^2} = \frac{n\rho_1^2 C_{\text{vdW}}^{(11)}}{24 a_o^2}. \tag{4.21}$$

where the coefficient $C_{\text{vdW}}^{(11)}$ is to be evaluated using Eq. (3.72). This procedure can be extended to provide an expression for γ_{12} from Eq. (4.19):

$$\gamma_{12} \simeq \gamma_1 + \gamma_2 - \frac{H_{12}}{12\pi a_o^2}$$

$$= \frac{n}{24 a_o^2}\left[\rho_1^2 C_{\text{vdW}}^{(11)} + \rho_2^2 C_{\text{vdW}}^{(22)} - 2\rho_1\rho_2 C_{\text{vdW}}^{(12)}\right] \tag{4.22}$$

Now both $C_{\text{vdW}}^{(11)}$ and $C_{\text{vdW}}^{(22)}$ must be evaluated using Eq. (3.72) while Eq. (3.71) must be used to evaluate $C_{\text{vdW}}^{(12)}$. This result is most valid when a solid interface is formed between two materials that interact predominantly by van der Waals forces alone.

4.3   The Derjaguin Approximation

Based on the above discussion, much can be learned by measuring the interaction potential energy (or equivalently the interaction force) between

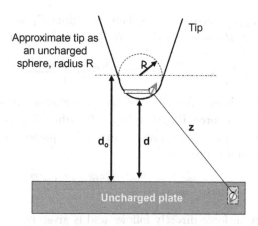

Fig. 4.8 A tip of radius R positioned a distance d above a flat uncharged substrate. vdW forces act between each atom in the tip and each atom in the substrate. The force on the tip can be inferred using the Derjaguin approximation.

two flat planes separated by a distance d. In practice, it is difficult to realize this parallel geometry for nanometer-size gaps since small tilt angles between the two planar surfaces are difficult to eliminate. It is much easier to measure the interaction between a sphere and a flat plane since the parallel geometric constraint is no longer required. It would be useful if we could relate the fundamental (but difficult to achieve) plane-plane interaction energy to the more easily accessible plane–sphere geometry illustrated in Fig. 4.8.

The Derajaguin approximation relates the force law, $F(d)$, between two curved surfaces having local radii R_1 and R_2 to the interaction potential energy per unit area, $U(d)/A$, between two planar surfaces [derjaguin34]. This approximation is very useful since it is usually easier to derive the interaction energy for two planar surfaces rather than for curved surfaces. The Derjaguin approximation can be formally stated as

$$F_{\text{curved}}(d) \simeq 2\pi R_{\text{eff}} \frac{U_{\text{plane}-\text{plane}}(d)}{A}, \tag{4.23}$$

where the effective radius of the two curved surfaces is defined by

$$\frac{1}{R_{\text{eff}}} = \frac{1}{R_1} + \frac{1}{R_2}. \tag{4.24}$$

Equation (4.23) is only valid when the radii R_1 and R_2 are much greater than the surface-to-surface separation between the two curved surfaces. As illustrated in Fig. 4.8, in most AFM experiments, the AFM tip can reasonably be approximated as a sphere of radius $R_1 = R$ while the flat plate has a radius $R_2 \to \infty$, so $R_{\text{eff}} = R$. This implies that

$$F_{\text{plane}-\text{sphere}}(d) \simeq 2\pi R \frac{U_{\text{plane}-\text{plane}}(d)}{A}. \tag{4.25}$$

This same result also follows from the pair-wise additive derivation of the Hamaker surface force. Based on Eq. (4.9), when $R \gg d$ the interaction energy between a sphere of radius R and a flat plane with a surface-to-surface separation d is given by

$$U_{\text{plane}-\text{sphere}}(d) = -\frac{\pi^2 \rho_1 \rho_2 C_{\text{vdW}}}{6} \frac{R}{d} \tag{4.26}$$

The interaction force directly follows and is given by

$$F_{\text{plane}-\text{sphere}}(d) = -\frac{\partial U_{\text{plane}-\text{sphere}}(d)}{\partial d} = -\frac{\pi^2 \rho_1 \rho_2 C_{\text{vdW}}}{6} \frac{R}{d^2} \tag{4.27}$$

The interaction potential energy per area A between two planes has also been derived and is given by Eq. (4.17)

$$\frac{U_{\text{plane}-\text{plane}}(d)}{A} = -\frac{\pi \rho_1 \rho_2 C_{\text{vdW}}}{6} \frac{1}{2d^2} \tag{4.28}$$

where the area A is defined in Fig. 4.7.

From Eqs. (4.27) and (4.28), it follows that the force acting on a sphere of radius R with a surface-to-surface separation d above a flat plane can be expressed in terms of the interaction energy per unit area for two flat planes:

$$F_{\text{plane}-\text{sphere}}(d) = 2\pi R \left(\frac{U_{\text{plane}-\text{plane}}(d)}{A} \right). \tag{4.29}$$

This result is the same as obtained by the Derjaguin approximation in Eq. (4.25).

These results can also be extended to learn more about the surface energies γ. When a sphere contacts a plane, then $d = a_o$, and using Eqs. (4.19) and (4.20), we have an expression for the force acting on the tip

$$F_{\text{plane}-\text{sphere}}(d) = 2\pi R \times \begin{cases} 2\gamma_1 & \text{if the two materials are identical} \\ \gamma_1 + \gamma_2 - \gamma_{12} & \text{if the two materials are different} \end{cases} \tag{4.30}$$

Equation (4.30) shows that the force acting on the tip at contact is a combination of geometry $(2\pi R)$ and material properties (γ).

Example 4.3: Estimate the force exerted by a clean Si tip of radius 5 nm when it contacts a clean, flat Si wafer. Assume the measurement is made in ultra-high vacuum so there is no native oxide layer (SiO_2) on the tip or wafer. Use a value of $\gamma = 850 \, mJ/m^2$ [keene93].

Since the tip and substrate are nominally the same material, the correct form of the Derajaguin approximation (Eq. (4.30)) is

$$F_{\text{plane-sphere}}(\text{contact}) = 2\pi R \times 2\gamma_1$$
$$= 6.28 \cdot 5 \times 10^{-9} m \cdot 2 \cdot 0.850 \, J/m^2$$
$$= 53 \times 10^{-9} N = 53 \, nN$$

The actual measured value could be different due to native oxide layers, surface roughness of the Si wafer and tip, and any uncontrolled and unwanted surface contamination from residual gases.

The Derajaguin approximation is considerably more powerful than the special case of the sphere–plane discussed above. If the force $F(z)$ acting between **two planes** separated by a distance z is known, then the force $F(d)$ acting between a flat plane and an object of **arbitrary shape** can be approximated by

$$F(d) = \int_d^\infty F(z)\frac{dA}{dz}dz \qquad (4.31)$$

where dA/dz describes how the cross-sectional area of the object with arbitrary shape changes with z.

Example 4.4: Actual cantilevers used in AFM experiments do not have spherical tips. The tip shape can be approximated by a cone with a smooth rounded end. Therefore consider the slightly more realistic model of a tip described by a circular parabaloid of revolution. What is the force between this tip and a flat plane when the surface-to-surface separation is d_o?

(Continued)

Example 4.4: (*Continued*)

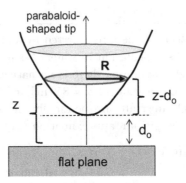

The apex of the parabaloid is a distance d_o above a flat plane as shown in the diagram. Assume the flat plane and parabaloid tip are comprised of identical atoms that interact by London dispersion forces governed by a Hamaker constant H.

The shape of the tip can be approximated by a circular parabaloid of revolution which is described by an equation of the form

$$z - d_o = \frac{x^2 + y^2}{a^2},$$

where d_o is the offset distance from the flat plane and the adjustable parameter "a" (with units of $nm^{1/2}$) controls the "sharpness" of the tip. A plot of this equation for $d_o = 2\,nm$ and a few values of "a" is given in the figure below. This model can be used in AFM to approximate the shape of a tip near its apex.

It follows that the radius $R(z)$ (for $z > d_o$) of a circular cross-section a distance $(z - d_o)$ above the tip apex is $R(z) = \sqrt{x^2 + y^2} = a\sqrt{(z - d_o)}$. The cross-sectional area A is then given by

$$A(z) = \pi R^2(z) = \pi a^2 (z - d_o).$$

The change in area with respect to z is $\frac{dA}{dz} = \pi a^2$.

(*Continued*)

Example 4.4: (*Continued*)

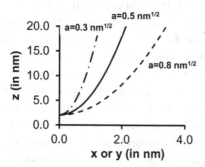

From Eq. (4.18a), we have

$$\frac{U_{\text{plane}-\text{plane}}(z)}{A} = -\frac{H}{12\pi z^2}$$

$$F(z) = -\frac{\partial}{\partial z}\left(-\frac{H}{12\pi z^2}\right) = -\frac{H}{6\pi z^3}$$

From Eq. (4.31)

$$F(d) = \int_d^\infty F(z)\frac{dA}{dz}dz = -\frac{H}{6\pi}\int_{d_o}^\infty \frac{1}{z^3}\pi a^2 dz = \frac{Ha^2}{6}\left[\frac{1}{2z^2}\right]_{d_o}^\infty$$

$$= -\frac{Ha^2}{12}\frac{1}{d_o^2}.$$

The Derjaguin approximation is generally valid so long as the interaction force does not change rapidly with the curvature of the tip.

4.4 Van der Waals Interactions from a Condensed Matter Perspective

The discussion in Chapter 3 of the vdWs interaction focused on gas phase atoms. It was assumed, for instance, that the atoms were mobile enough to justify a Boltzmann average over all relevant orientations. When gas phase atoms condense into the solid state, this positional averaging may not be relevant. As a consequence, the evaluation of individual contributions to C_{vdW} found in Table 3.3 must be reconsidered. By the 1950s, it also became clear that Hamaker's pair-wise summation method to calculate vdWs dispersion forces could be improved by focusing squarely on the electrodynamic charge fluctuations of atoms in the condensed state. Following this approach to a logical end requires a deep understanding of the electromagnetic properties of condensed matter with particular emphasis on dielectric response functions and the resulting electromagnetic adsorption spectrum, topics not often studied by most students using an AFM. A fundamental assumption underlying this approach is that the frequency at which charges spontaneously fluctuate in solids roughly coincides with the frequency at which atoms resonantly adsorb energy from incident electromagnetic waves, a principle that is embedded in the fluctuation-dissipation theorem studied in statistical thermodynamics. This approach to understanding the vdW force shifts the emphasis away from the properties of isolated atoms and focuses more on the properties relevant to the condensed state of matter. Basically, the frequency-dependent electromagnetic adsorption spectrum of the condensed state itself is now used to directly calculate the dispersion force between macroscopic bodies.

Classical physics is no longer a reliable guide when discussing quantum processes and quantum electrodynamics which were developed to explain the emission of radiation when an atom undergoes a transition from one quantum state to another. One prediction of quantum electrodynamics is that electric and magnetic fields constantly fluctuate about their classical values, even in the absence of electromagnetic waves or steady $\vec{E}$ and $\vec{B}$ fields. According to Heisenberg's uncertainty principle, the energy contained in these vacuum fluctuations (ΔE) does not violate energy conservation if the fluctuations persist for a short enough time (Δt) so that

$$\Delta E \Delta t \geq \frac{\hbar}{2}. \tag{4.32}$$

These vacuum fluctuations can be viewed as virtual photons and photons exert a radiation pressure when interacting with a solid. This leads to a force between two slabs of material separated by a distance d. A schematic of the physical situation is sketched in Fig. 4.9. Outside the slabs, all photon wavelengths are possible, while between the two slabs, only discrete wavelengths are allowed due to the quantization imposed by the separation distance d. A net inward force develops.

What charge fluctuations might be important in this context? In principle, any charge fluctuation could be important whether it is due to electrons moving around atoms, dipoles that either vibrate or rotate, or mobile electrons in metals — basically any charge that can respond to an applied

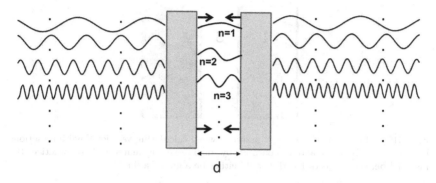

Fig. 4.9 Fluctuations in the electromagnetic (EM) fields near two solid blocks of material separated by a distance d. To the left and right of the two blocks, EM fields of all frequency and wavelength are possible. Between the blocks, only quantized EM fields (modes) that satisfy boundary conditions can exist. The reduced number of EM modes between the blocks produces a net inward force.

electric field. This means there can be many parallel contributions to the dielectric function that relates an induced dipole moment p to the local electric field E_{loc} at any point in the solid. Understanding this approach is admittedly difficult and the calculation relies heavily on quantum field theory, a topic that lies well beyond the interest of most students of AFM. Nonetheless, it is important to appreciate some of the important results that have followed from this way of thinking.

A general theory to calculate the Hamaker constant directly was developed by Dzyaloshinskii, Lifshitz and Pitaevskii (DLP) in 1959 [dzyaloshinskii61]. Their theory assumes knowledge of the macroscopic dielectric functions of the interacting objects. Most notably, the assumption of pairwise additivity discussed in Sec. 4.1 need not be made. The correlation of spontaneous electric charge fluctuations between closely-spaced neighboring objects produces a non-zero force resulting from the fluctuating electromagnetic fields that exist. Implicit in this theory is the assumption that the gap between the two objects must be larger than molecular dimensions. The interacting objects are treated as continuous and the force between the macroscopic bodies requires knowledge of the frequency dependence of the dielectric functions of the interacting materials over a wide range of frequencies. In essence, the dielectric function provides the appropriate volume-averaging algorithm that is approximated by Hamaker's pairwise additive approach.

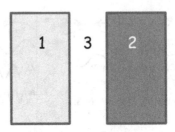

Fig. 4.10   The standard geometry assumed when calculating van der Waal interactions between two objects 1 and 2 separated by an intervening material 3. In practice, the material between objects 1 and 2 could either be a gas or a liquid.

The main result from the DLP theory is an equation for the energy per unit area of object 1 interacting with object 2 when the two are separated by material 3 all held at a temperature T (see Fig. 4.10). Objects 1 and 2 are considered as two infinite half slabs and the third material 3 is assumed to fill the gap d between objects 1 and 2. In the language of the discussion found in Sec. 4.1, the interaction can be described in terms of a Hamaker constant H_{132}.

If the gap d is small (non-retarded limit) and if materials 1 and 2 are non-magnetic and if the dielectric properties of material 1 and 2 are not too different, then a simplified form of the DLP theory gives [parsegian06]

$$\frac{U(d)}{A} \simeq -\frac{k_B T}{8\pi d^2} \sum_{n=0}^{\infty}{}' \left[\frac{\varepsilon_1(i\nu_n) - \varepsilon_3(i\nu_n)}{\varepsilon_1(i\nu_n) + \varepsilon_3(i\nu_n)}\right] \left[\frac{\varepsilon_2(i\nu_n) - \varepsilon_3(i\nu_n)}{\varepsilon_2(i\nu_n) + \varepsilon_3(i\nu_n)}\right] \qquad (4.33)$$

where $U(d)/A$ is the interaction energy per unit area, $\varepsilon_1(i\nu_n)$ and $\varepsilon_2(i\nu_n)$ are the frequency dependent permittivity of materials 1 and 2 evaluated at imaginary angular frequencies $i\nu_n$, $\varepsilon_3(i\nu_n)$ is the frequency dependent dielectric function of the surrounding medium. If the two materials are in vacuum, $\varepsilon_3(i\nu) = 1$ in Eq. (4.33). The $'$ on the sum indicates that the $n = 0$ term in the summation must be multiplied by $1/2$. The frequencies are known as the Matsubara frequencies

$$\nu_n = \left(\frac{2\pi k_B T}{\hbar}\right) n \qquad (4.34)$$

where n is now a positive integer and $\hbar = 1.054 \times 10^{-34}\,\mathrm{J \cdot s}$ is Planck's constant divided by 2π. In Eq. (4.33), the dielectric functions are evaluated at imaginary frequencies, an unfortunate choice of words but one that is used so widely in the theoretical literature that we dare not change it. The basic idea behind an imaginary frequency is based on the well-known convention that for a given angular frequency ω, in-phase and out-of-phase sinusoidal oscillations can be conveniently represented by Euler's identity: $e^{i\omega t} = \cos\omega t + i\sin\omega t$. Now if a function is evaluated when $\omega = i\nu$ (i.e., at an imaginary frequency), then $e^{i\omega t} = e^{i(i\nu)t} = e^{-\nu t}$. The implication is that any factor that oscillated with time will now be converted to a factor that decays exponentially in time. This provides a convenient way to describe how a fluctuating charge might return to its original, undisturbed position.

Equation (4.33) contains the product of two terms often called the spectral mismatch functions and it is the product of these two terms that determines the interaction between objects 1 and 2. It is also evident that if the dielectric function of the surrounding medium 3 happens to equal the dielectric function of either one of the two objects *at all frequencies*, than $H_{132} = 0$ and no interaction force will result.

The prominent role of the dielectric functions in Eq. (4.33) emphasizes the importance in the spectral mismatch between the interacting objects, a result that can be linked back to the different polarizabilities used in molecule–molecule formulations of the dispersion interaction discussed in Chapter 2. The dielectric function (which is different from the static dielectric constant) essentially determines the ability of a material to neutralize a sinusoidally varying applied electric field that varies with time at an angular frequency ω. Depending on the values of the spectral mismatch terms in Eq. (4.33), the net interaction can be attractive, repulsive or even zero. As evident in Eq. (4.33), the frequency dependence is reduced to an infinite sum over a discrete set of frequencies, somewhat reminiscent of Planck's treatment of blackbody radiation modes in hollow cavities that dates to the early 1900s. However, in this case the final result is derived from the analysis of complex functions. The sum is often replaced by an integral with an integrand containing singularities (poles) that must be evaluated according to Cauchy's residue theorem.

If one chooses to focus on the Hamaker constant between objects 1 and 2 separated by a distance d filled with an intermediate material 3, Eq. (4.33) can be rewritten to give an expression for H_{132} [parsegian06]

$$\frac{U(d)}{A} = -\frac{H_{132}}{12\pi d^2} \simeq -\frac{k_B T}{8\pi d^2} \sum_{n=0}^{\infty}{}' \left[\frac{\varepsilon_1(i\nu_n) - \varepsilon_3(i\nu_n)}{\varepsilon_1(i\nu_n) + \varepsilon_3(i\nu_n)}\right] \left[\frac{\varepsilon_2(i\nu_n) - \varepsilon_3(i\nu_n)}{\varepsilon_2(i\nu_n) + \varepsilon_3(i\nu_n)}\right].$$

$$(4.35)$$

This leads to the following formula for H_{132}

$$H_{132} = \frac{3}{2}k_B T \sum_{n=0}^{\infty}{}' \left[\frac{\varepsilon_1(i\nu_n) - \varepsilon_3(i\nu_n)}{\varepsilon_1(i\nu_n) + \varepsilon_3(i\nu_n)}\right] \left[\frac{\varepsilon_2(i\nu_n) - \varepsilon_3(i\nu_n)}{\varepsilon_2(i\nu_n) + \varepsilon_3(i\nu_n)}\right]. \qquad (4.36)$$

The result is interesting because it can be related to the previous discussion of Hamaker constants:

(1) The Hamaker constant is proportional to $k_B T$, indicating the role that thermal motion plays in the charge fluctuation process.

(2) The Matsubara frequencies ν_n in Eq. (4.36) are quantized ($n = 1, 2, \ldots$). Hence the energy of the photons ($\hbar \nu_n$) are also quantized. The energy of the contributing photons in the summation indicates they are integer multiples of thermal energies $k_B T$.

(3) At room temperature, the $n = 1$ term corresponds to frequencies in the infrared. Higher values of n indicate the importance of photons in the visible and UV regions of the electromagnetic spectrum.

(4) It can be shown [parsegian06] that the $n = 0$ (zero frequency) term in Eq. (4.36) reproduces the Keesom and Debye interactions discussed in Chapter 3. For strongly polar materials, the zero-frequency term may be significant. Furthermore, it follows that the terms containing the sum over frequencies ($n = 1, 2, \ldots$) in Eq. (4.36) must account for the correlated London dispersion interaction between fluctuating charges.

Although very general, the theory outlined above is difficult to apply without making some simplifying assumptions.

Historically, the lack of data for the full spectral optical properties of materials has hindered the use of the DLP model, but this is rapidly changing as measurements of vacuum ultra violet (VUV) optical reflectivity are becoming more available [weaver81]. With modern light sources like synchrotron radiation, these frequency dependent dielectric functions can be experimentally measured and tabulated over a wide range of frequencies to serve as input parameters to evaluate relevant vdW interactions from first principles.

It is useful to discuss the dielectric function in a bit more detail to provide further insight and to allow a numerical evaluation of the Hamaker constant using the DLP theory. The following discussion is probably of most interest to students in condensed matter physics and summarizes a vast amount of literature in an abbreviated fashion. It does provide some context for the evaluation of Eq. (4.36), an unlikely occurrence today that will likely become a common trend in the future.

4.5   The Frequency Dependent Dielectric Function

To use the DLP model, the dielectric function for particular materials must be measured over a wide frequency range (from $\sim 0.01\,\text{eV}$ to $\sim 50\,\text{eV}$). It is worthwhile to briefly review the physics underlying the frequency dependent dielectric function.

In the most general terms, to screen electric fields, a material must contain localized charge that can be displaced by an electric field. Charge displacement (also referred to as polarization) provides an inherent energy loss mechanism that attenuates an applied electric field. Since any charge displacement will roughly follow the time dependence of an applied electric field, it is customary to define a complex permittivity $\varepsilon(\omega)$ to describe the in-phase, out-of-phase response of the material system.

$$\varepsilon(\omega) = \varepsilon'(\omega) + i\varepsilon''(\omega) \tag{4.37}$$

The real part of $\varepsilon(\omega)$ reflects the energy stored while the imaginary part of $\varepsilon(\omega)$ reflects the energy lost. If the imaginary part of the dielectric function is close to zero in any region of the electromagnetic spectrum, the material will be transparent to EM radiation over that wavelength range. CGS units are typically used in this discussion, so technically, $\varepsilon(\omega)$ is the relative permittivity, the quantity κ used in the previous chapter. To obtain the dielectric function in SI units, you need to multiple by ε_o.

The dielectric function in turn is related to the complex refractive index of a material according to

$$\varepsilon(\omega) = \varepsilon'(\omega) + i\varepsilon''(\omega) = (n(\omega) + ik(\omega))^2 \tag{4.38}$$

where

$$n(\omega) = \sqrt{\frac{\varepsilon' + \sqrt{(\varepsilon'^2 + \varepsilon''^2)}}{2}} \tag{4.39}$$

and

$$k(\omega) = \sqrt{\frac{-\varepsilon' + \sqrt{(\varepsilon'^2 + \varepsilon''^2)}}{2}} \tag{4.40}$$

Thus measurements of n and k as a function of the frequency ω of an incident electromagnetic wave are often used to define ε.

In general terms, the polarization mechanism(s) that occur in materials depend on frequency, temperature, and composition and the structure of the material under study. The variation in the complex permittivity $\varepsilon''(\omega)$ as a function of ω must reflect the hidden details that are relevant in the solid state.

As an example, at low frequencies, the displacement of charge in most materials is dominated by space charge separation (bulk polarization) that occurs primarily at interfaces. This topic is usually covered in most introductory physics courses on electricity and magnetism. As the frequency of an applied electric field increases, other mechanisms like permanent dipolar orientation or rotation (microwave), resonant molecular vibration or optical lattice vibrations (infrared), and atomic polarization or electron displacement (visible) come into play. Due to absorption, measured dielectric functions for real materials exhibit a complicated behavior that can be a highly non-monotonic function of frequency, making the analytical evaluation of Eq. (4.36) problematic.

The dielectric function appearing in Eq. (4.36) at imaginary frequencies is defined as

$$\varepsilon(i\nu_n) = 1 + \frac{2}{\pi} \int_0^\infty \frac{\omega \varepsilon''(\omega)}{\omega^2 + \nu_n^2} d\omega. \tag{4.41}$$

Note that the multiplicative term $\omega/(\omega^2 + \nu_n^2)$ in Eq. (4.41) acts somewhat as filter that effectively selects only the values of $\varepsilon''(\omega)$ near the frequency ν_n. For $\nu_n = 0$, $\varepsilon(i\nu)$ should reproduce the dielectric constant κ. If a material is transparent in the IR, visible or near UV, then $\varepsilon(i\nu)$ should exhibit a broad plateau with a value close to the squared index of refraction n^2 over the transparent frequency range.

Are there any simple predictions for the frequency dependence of the dielectric function in a homogeneous, isotropic material? Various classic analytical models have been developed and elaborated to describe the polarization of bound charge (Lorentz), the polarization of free charges in metals (Drude), and the polarization when dipolar effects are important (Debye). Although these models are useful to build intuition, the availability of modern computational tools makes the evaluation of Eqs. (4.36) and (4.41) reasonably straightforward, if the imaginary part of the dielectric function for a given material is known [parsegian06].

Even though measured values of $\varepsilon''(\omega)$ are full of interesting structure that mirrors the different loss mechanisms inherent in the material of interest, the $\varepsilon(i\nu_n)$ as defined by Eq. (4.41) are a rather featureless function of ν_n due to the action of the integration. Increasingly, reliable information about $\varepsilon''(\omega)$ can be found in the original literature or by consulting reference data tables. This information will be increasingly used to provide reliable estimates of the Hamaker constants in AFM experiments rather than relying on dubious assumptions based on overly simplistic models.

4.6 Chapter Summary

The vdW interactions between individual atoms discussed in Chapter 3 are used to calculate the interaction between objects containing many atoms. Two geometries are of particular interest to AFM applications: a block oriented parallel to an infinite flat surface and a sphere positioned above a flat surface. In both cases, the surface-to-surface separation is d.

The results for these two important geometries are collected for future reference in Fig. 4.11.

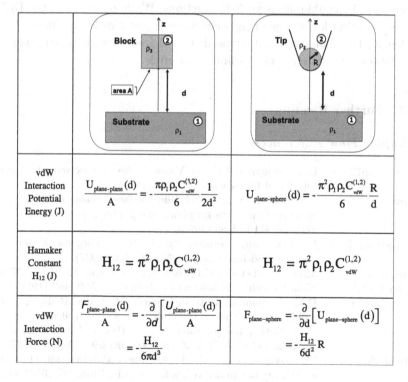

vdW Interaction Potential Energy (J)	$\dfrac{U_{\text{plane-plane}}(d)}{A} = -\dfrac{\pi\rho_1\rho_2 C_{\text{vdW}}^{(1,2)}}{6}\dfrac{1}{2d^2}$	$U_{\text{plane-sphere}}(d) = -\dfrac{\pi^2\rho_1\rho_2 C_{\text{vdW}}^{(1,2)}}{6}\dfrac{R}{d}$
Hamaker Constant H_{12} (J)	$H_{12} = \pi^2\rho_1\rho_2 C_{\text{vdW}}^{(1,2)}$	$H_{12} = \pi^2\rho_1\rho_2 C_{\text{vdW}}^{(1,2)}$
vdW Interaction Force (N)	$\dfrac{F_{\text{plane-plane}}(d)}{A} = -\dfrac{\partial}{\partial d}\left[\dfrac{U_{\text{plane-plane}}(d)}{A}\right]$ $= -\dfrac{H_{12}}{6\pi d^3}$	$F_{\text{plane-sphere}} = -\dfrac{\partial}{\partial d}\left[U_{\text{plane-sphere}}(d)\right]$ $= -\dfrac{H_{12}}{6d^2}R$

Fig. 4.11   A summary of the important results discussed in Chapter 4.

Using Hamaker's pair-wise additive theory, analytical formulas for the interaction potential energy and interaction force were derived.

In the limit when any two objects are separated by an atomic distance, Hamaker's results can be used to estimate γ, the surface energy of a solid. Derajaguin's approximation is also discussed since it provides a framework to extend the very basic result obtained for the case of two parallel flat surfaces to more relevant geometries. Lastly, the Dzyaloshinskii, Lifshitz and Pitaevskii (DLP) theory to calculate inter-molecular interactions between solid objects is reviewed. The DLP theory is very general and relies on an analytical description of the fluctuating electromagnetic fields that are present near any object formed by a collection of atoms. This theory removes the constraint of pair-wise additive interactions required in Hamaker's theory by relying on the frequency dependence of the dielectric function $\varepsilon(\omega)$ required to describe the response of a solid material to a time-varying electromagnetic field. The final result reveals the role played by the dielectric properties of an intervening medium, a role that can reduce and even cancel the ubiquitous vdW interactions. While more exact, the DLP theory requires knowledge of the material dependent dielectric function to calculate the Hamaker constant needed to characterize the magnitude of the interaction between two dissimilar materials.

4.7   Further Reading

Chapter Four References:

[bargeman72]   D. Bargeman and F. van Voorst Vader, J. Electroanal. Chem. Interfacial Electrochem. **37**, 45 (1972).

[bauer96]   J. Bauer, G. Drescher, M. Illig, Surface tension, adhesion and wetting of materials for photolithographic process, J. Vac. Sci. Technol. **B14**, 2485 (1996).

[bergstrom97]   L. Bergstrom, Hamaker Constants of Inorganic Materials, Adv. Colloid Interface Sci. **70**, 125–169 (1997).

[croucher77]   M.D. Croucher and M.L. Hair, J. Phys. Chem. **81**, 1631 (1977).

[croucher81]   M.D. Croucher, Colloid and Polymer Sci. **259**, 462 (1981).

[derjaguin34]   B.V. Derjaguin, Untersuchungen über die Reibung und Adhäsion, IV. Theorie des Anhaftens kleiner Teilchen [Analysis of friction and adhesion, IV. The theory of the adhesion of small particles]. Kolloid Z. (in German) **69**, 155–164 (1934).

[dzyaloshinskii61]   I.E. Dzyaloshinskii, E.M. Lifshitz and L.P. Pitaevskii, The general theory of van der Waals forces, Adv. Phys. **10**, 165 (1961).

[hamaker37]   H.C. Hamaker, Physica **4**, 1058 (1937).

[israelachvili98]   J.N. Israelachvili, *Intermolecular and surface forces*. Academic Press, New York (1998).

[jasper72]   J.J. Jasper in J. Phys. Chem. Ref. Data **1**, 841–980 (1972).

[keene93]   B.J. Keene, Review of data for the surface tension of pure metals, Int. Maters. Rev. **38**, 157–92 (1993).

[kittel86]   C. Kittel, *Introduction to Solid State Physics*, 6th ed., Wiley and Sons, New York (1986).

[leite12]   F.L. Leite, C.C. Bueno, A.L. Da Róz, E.C. Ziemath and O.N. Oliveira Jr., Theoretical Models for Surface Forces and Adhesion and Their Measurement Using Atomic Force Microscopy, Int. J. Mol. Sci., **13**, 12773–12856 (2012).

[parsegian06]   V. Adrian Parsegian, *van der Waals Forces*, Cambridge University Press, New York (2006).

[visser72]   J. Visser, On Hamaker Constants: A Comparison between Hamaker Constants and Lifshitz-van der Waals Constants, Adv. Colloid Interface Sci. **3**, 331–363 (1972). (Provides tables of Hamaker constants for different materials).

[weaver81]   J.H. Weaver, C. Krafka, D.W. Lynch and E.E. Koch, *Optical Properties of Metals, Pts. I and II* in Physik Daten Nr. 18-1 and Nr. 18-2, Fachinformationszentrum Energie, Physik, Mathematik GmbH, Karlsruhe, Germany (1981). (Comprehensive data tables for the complex dielectric function of most chemical elements.)

Further reading:

[argento96]   C. Argento and R.H. French, Parametric tip model and force distance relation for Hamaker constant determination from atomic force microscopy, J. Appl. Phys. **80**, 6081–90 (1996).

[burns85]   Gerald Burns, *Solid State Physics*, Academic Press, New York (1985).

[butt03]   H.-J. Butt, K. Graf, M. Kappl,, *Physics and Chemistry of Interfaces*, Wiley-VCH Verlag & Co. KGaA, Weinheim, Germany (2003).

[capella99]   B. Cappella, G. Dietler, Force-distance curves by atomic force microscopy, Surf. Sci. Rep. **34**, 1–104 (1999).

[fernandez-varea00]   J.M. Fernandez-Varea and R. Garcia-Molina, Hamaker Constants of Systems Involving Water Obtained from a Dielectric Function that Fulfills the f-Sum Rule, J. Colloid Inter. Sci. **231**, 394 (2000).

[hiemenz97]   P.C. Hiemenz and R. Rajagopalan, *Principles of Colloid and Surface Chemistry*, 3rd ed., Marcel Dekker, New York (1997).

[mcclellan63]   A.L. McClellan, *Tables of Experimental Dipole Moments*, W.H. Freeman, San Francisco CA (1963).

[oss88]   C.J. van Oss, M.K. Chaudhury, R.J. Good, Interfacial
                     Lifshitz-van der Waals and Polar Interactions in Macroscopic
                     Systems, Chem Rev. 88, 927–41 (1988).
[richmond75]   P. Richmond, The theory and calculation of van der Waals
                     forces, in Colloid Science, Vol. 2, ed. D.H. Everett, The
                     Chemical Society, Burlington House, London (1975).
[rigby86]   M. Rigby, E.B. Smith, W.A. Wakeham, G.C. Maitland, *The
                     Forces Between Molecules*, Clarendon Press, Oxford (1986).

4.8 Problems

1. What force does a neutral Au atom experience when suspended 1 nm above a plane of Au atoms? Assume the Hamaker constant for Au is 300×10^{-21} J.

2. After removing a Si wafer from etch, it rapidly develops a thin layer of oxide under ambient conditions. Examination of the wafer with an optical microscope reveals it has unfortunately become contaminated with small SiO_2 particles. Is it easier to remove the SiO_2 particles by directing a stream of air onto the wafer or by ultrasonically cleaning the wafer in a water bath? The Hamaker constant for SiO_2 interacting with SiO_2 in air is about 6.5×10^{-20} J. For SiO_2 interacting with SiO_2 in H_2O, the Hamaker constant decreases to a value of about 0.9×10^{-20} J.

3. A small crystalline grain is deposited onto a flat substrate. The grain has a cubic shape with sides of length 50 nm. Assume that γ for this system has a value of 0.25 eV per 5 square nanometers (mixed units). What is the van der Waals force holding the crystalline grain to the substrate?

4. Estimate the van der Waals force holding a spherical polymer particle to a flat substrate. Assume the sphere has a radius of 1 μm and that the Hamaker constant is $H = 8.5 \times 10^{-20}$ J.

5. Order of magnitude estimates for Hamaker constants can be found throughout the AFM literature. It is common to find statements that for non-conducting (i.e., non-metallic) solids interacting in vacuum or air, the Hamaker constant is typically in the range $H = (5-10) \times 10^{-20}$ J. For interactions in a liquid such as water, the Hamaker constant is typically one order of magnitude smaller, in the range $H = (0.5-1.5) \times 10^{-20}$ J. Explain why this is a reasonable claim.

6. Surprisingly, van der Waal forces can be quite large if the separation between objects with atomically smooth surfaces becomes very small. Suppose the ends of two Au rods with a diameter of 1 mm are polished to atomic flatness and then pushed together. How much force will be required to pull the interface apart? The Hamaker constant for Au is about 0.30 aJ. If the separation force is so large, why do we need to solder two wires together?

7. Using the Derjaguin approximation, what is the force of attraction between two electrically neutral polystyrene spheres of 1 μm radius when they are separated by a distance of 10 nm in air? The Hamaker constant for polystyrene acting across air is about 7×10^{-20} J. If equal numbers of electrons are added to each sphere, how many electrons would be needed to cancel the attractive van der Waals force?

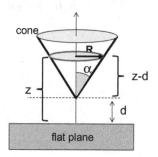

8. Using the Derjaguin approximation, estimate the force between a conical tip of half angle α when it is located a distance d from a flat substrate.

9. You can show that two spheres of radii R_1 and R_2 separated by a surface-to-surface separation distance d and interacting by van der Waals forces have an interaction potential energy given by

$$U(d) = -\frac{H}{6} \frac{R_1 R_2}{R_1 + R_2} \frac{1}{d}.$$

If the radii of both spheres are increased by a factor η ($\eta > 1$) and the surface-to-surface separation distance d is increased by the same factor η, how does the interaction potential energy change? What does this result imply for the van der Waals force acting between two spheres of radius 1 nm separated by 5 nm compared to two spheres of radius 100 nm separated by 500 nm?

10. Consider a sphere of radius R interacting with a large flat plane by van der Waals forces. Let the surface-to-surface separation distance between

these two objects equal d. Under what geometric constraints will the vdW interaction be dominant over thermal energies $k_B T$? In other words, what condition must be imposed on the ratio of R/d so that the vdW interaction will be greater than thermal energies?

11. Consider dust particles of diameter $\sim 2\,\mu m$ floating in air. How close must two particles (assume they are composed of SiO_2) approach each other so that they stick together via vdW forces? For this to happen, assume the van der Waals attraction must equal $\sim 5\,k_B T$. The Hamaker constant for SiO_2 interacting with SiO_2 across air is about $6 \times 10^{-20}\,J$. Approximate the van der Waals interaction energy using the Hamaker result for two spheres of radii R_1 and R_2 separated by a surface-to-surface separation distance d

$$U(d) = -\frac{H}{6}\frac{R_1 R_2}{R_1 + R_2}\frac{1}{d}.$$

Chapter 5

When the Tip Contacts
the Substrate: Contact Mechanics

In the previous chapter, the force law acting between objects with micron-size dimensions and separated by a few tens of nanometers was discussed. When the separation distance between two objects becomes small, on the order of an interatomic distance, the two solids contact each other and a strong repulsive force limits the penetration of one object into the other. A consequence of this repulsive force is a deformation in both the tip and the substrate. A quantitative understanding of this deformation requires an understanding of various contact mechanic models that span the range from a macroscopic continuum to the microscopic atomic length scale. To use an AFM intelligently, you must be aware of simple models used to estimate these deformations and you must understand what material properties control the deformations. In general terms, these topics are typically of most interest to students in materials science and mechanical engineering.

From the outset, a fundamental dilemma emerges. How should we consider the tip when modeling the tip–substrate interaction (see Fig. 5.1)? Specifically, does the tip end in one atom, or can it be represented by a smooth sphere of radius R_{tip}? Clearly the former view is more correct, but it presents a hopelessly intractable situation, since the position and chemical identity of each atom in the tip and each atom in the substrate then needs to be specified. A way forward is to realize that if the contact area between tip and substrate involves, say, hundreds of atoms, the description of the net repulsive force might be well approximated by continuum elasticity models.

Applying continuum models at the nanoscale can be risky since mechanical behavior could be quite different from that expected at the macroscale.

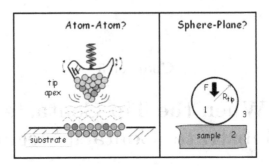

Fig. 5.1   How should the tip–substrate interaction be viewed when the tip is in close proximity to the substrate? Two possibilities are suggested, one which emphasizes the individual atom–atom interaction and a second which relies on the bulk properties of materials.

Although continuum models are probably accurate when sample dimensions are above ∼50 nm, the large surface to volume ratios at smaller sizes increasingly controls deformation properties. Certainly at nanometer length scales, surface effects become dominant and can significantly modify both the behavior and elasticity values expected from macroscopic experiments.

In what follows, relevant results are reviewed that describe the mechanical deformation of solids, with a focus on those models that are often used to interpret data from AFM experiments. Since AFMs are used to investigate a wide variety of different materials, we emphasize general trends spanning different classes of materials.

5.1   Elasticity of Materials

As a general introduction, it is important to realize there are three deformations that occur when a solid block of material with a cross-sectional area A is subjected to an external force.

As shown in Fig. 5.2, these deformations can be categorized as

(i) an elastic elongation that results when an outward normal force F_n (normal stress $= \sigma_n = F_n/A$) perpendicular to the face of a block,

(ii) an elastic shear deformation that results when a shear force F_t (shear stress $= \sigma_t = F_t/A$) is applied along the edge of a block causing an angle α to develop between the lower fixed face and the moveable upper face, and

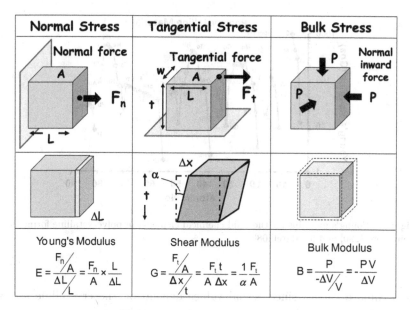

Normal Stress	Tangential Stress	Bulk Stress
Young's Modulus $$E = \dfrac{F_n/A}{\Delta L/L} = \dfrac{F_n}{A} \times \dfrac{L}{\Delta L}$$	Shear Modulus $$G = \dfrac{F_t/A}{\Delta x/t} = \dfrac{F_t\, t}{A\, \Delta x} = \dfrac{1}{\alpha}\dfrac{F_t}{A}$$	Bulk Modulus $$B = \dfrac{P}{-\Delta V/V} = -\dfrac{PV}{\Delta V}$$

Fig. 5.2 A schematic diagram illustrating three important deformations that occur in solids. Young's modulus (E), the shear modulus (G) and the bulk modulus (B) characterizing each of these deformations are defined.

(iii) an elastic compression that results when a hydrostatic pressure P produces an inward normal stress ($-\sigma_n$) to all faces of a block.

From the measured deformations ΔL, Δx and ΔV, three material-dependent coefficients (elastic moduli) can be defined as

$$E = \frac{\frac{F_n}{A}}{\frac{\Delta L}{L}} \quad G = \frac{\frac{F_t}{A}}{\frac{\Delta x}{L}} \quad B = \frac{P}{\frac{-\Delta V}{V}} \tag{5.1}$$

where E is Young's modulus, G is the shear modulus and B is the bulk modulus. Since the denominators in Eq. (5.1) have no units, all three quantities have units of pressure (N/m^2 or Pa).

5.1.1 Young's modulus

As we shall see, within the context of AFM, the most relevant elastic modulus is E, Young's modulus. Knowledge of this quantity for both the tip and substrate is required for quantitative interpretation of AFM data. Numerical values for Young's modulus are usually considered as

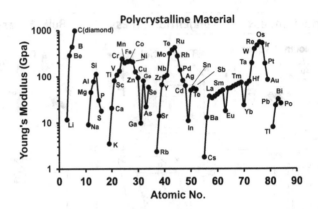

Fig. 5.3　Young's modulus for the solid chemical elements in polycrystalline form. Data from [gorecki80] and [cardarelli08].

approximate values, since actual values depend on the crystalline structure, orientation, and prior treatment of a particular sample.

Values for the Young's modulus for the solid chemical elements in polycrystalline form are given in Fig. 5.3.

As is evident from this survey chart, most elements in their polycrystalline form have a modulus ranging between 100 and 300 GPa.

We will not discuss in any detail the modifications to elasticity theory when a material is highly crystalline. Suffice it to say that the elastic properties for crystalline materials become considerably more complicated, since anisotropy in crystalline atomic structure produces different deformations when the same force is applied along different directions. This anisotropy, not readily apparent in bulk polycrystalline materials, becomes more important as size shrinks to the nanoscale where the atomic structure dominates and anisotropy in elasticity should play a more prominent role. This expectation foreshadows many possibilities, especially when conducting AFM experiments on nanometer size objects.

It is generally recognized that solids can be categorized into broad material classes: ceramics and glasses, metals and alloys, composites, polymers, elastomers (rubbers), wood products and foams. Since AFM is capable of scanning materials chosen from any of these material classes, it is useful to develop an intuition regarding values for Young's modulus for each material class. The chart in Fig. 5.4 plots the typical values of Young's modulus for these various material classes and illustrates how the modulus can range

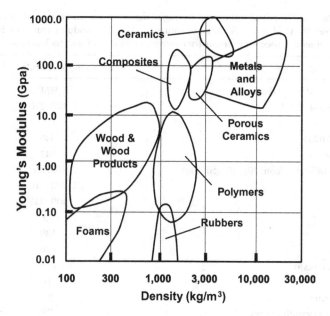

Fig. 5.4 The correlation between Young's modulus and density spanning a wide class of different materials [ashby89].

over ~5 orders of magnitude (10^5) while the material density changes by only a factor of ~300.

For reference, Table 5.1 lists values for the elastic moduli of various materials of general interest to those interested in AFM.

5.1.2 *Poisson's ratio*

When a solid is subjected to an outward tensile force, the elongation that results along the direction of the force is accompanied by a lateral contraction in the dimensions of a plane *perpendicular* to the applied force.

As shown in Fig. 5.5, the original cross-sectional area of dimensions $w \times d$ is deformed into a cross-sectional area of dimension $(w - \Delta w) \times (d - \Delta d)$. For an isotropic material undergoing an elastic deformation, the fractional changes in d and w are the same and measurements of these deformations allows us to define ν, the Poisson ratio, as

$$\nu = \frac{-\left(\frac{\Delta t}{t}\right)}{\frac{\Delta L}{L}} = \frac{-\left(\frac{\Delta w}{w}\right)}{\frac{\Delta L}{L}} = -\left(\frac{\Delta t}{\Delta L}\right)\left(\frac{L}{t}\right) = -\left(\frac{\Delta w}{\Delta L}\right)\left(\frac{L}{w}\right). \quad (5.2)$$

Table 5.1 Nominal values of Young's modulus for materials of general interest to AFM users. A range of values for Young's modulus can arise from variations in processing, impurity content, or measurement techniques.

Material	E, Young's Modulus (GPa)
Polycrystalline diamond	974
Stainless Steel	190–210
Ti	116
Al (7075)	72
Invar 36	147
Crystalline silicon: Si$\langle 110 \rangle$: Si$\langle 100 \rangle$	169:130
Si_3N_4	280–310
SiC	400–445
Au	79
Cu	130
W	400
Super Invar	148
Pyrex	61
Zerodur	91
SiO_2(crystalline quartz)	72
Sapphire (Al_2O_3)	345
SiO_2 (oxide film)	66
HOPG	36 ($\perp$ to basal plane)
Muscovite Mica	48 ($\perp$ to basal plane)
Polypropylene	0.9
Polycarbonate	2.6

The Poisson ratio ν is an important elastic parameter required to quantitatively interpret AFM data. As defined in Eq. (5.2), ν relates strain in one direction (x, also called longitudinal strain in the direction of the applied stress) to strain in a perpendicular direction (y or z, also called transverse strain at right angles to the applied stress).

For an object with free surfaces with no constraints, the theory of isotropic linear elasticity restricts the Poisson ratio to the range from -1 to $\frac{1}{2}$. A graph that summarizes measured values of ν for the solid chemical elements is given in Fig. 5.6.

From this chart, it is evident that many of the chemical elements that are solid near room temperature have values of ν close to 0.3. When $L \cong w \cong t$, then Eq. (5.2) indicates that Poisson's ratio measures the **ratio** of the distortion in a direction perpendicular to an applied force to the distortion in a direction parallel to the force.

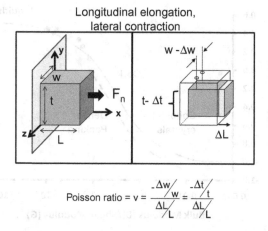

Fig. 5.5 An illustration of the deformation caused by a force applied perpendicular to the face of a material with a rectangular cross-section. The cross-sectional area of the cube before deformation is $t \times w$. After F_n is applied, the material elongates by an amount ΔL and the cross-sectional area change to $(t - \Delta t) \times (w - \Delta w)$.

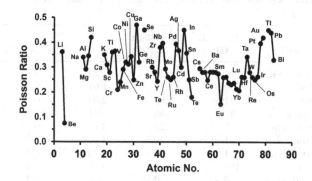

Fig. 5.6 Poisson's ratio for the solid chemical elements. Data from [cardarelli08].

Poisson's ratio also offers a fundamental metric that measures a material's resistance to volume alteration (measured by the bulk modulus B) compared to shear distortion (measured by the shear modulus G) when strained elastically. Since the AFM is often used to explore a wide variety of different samples, it is important to realize that different material classes exhibit systematic variations in ν as a function of the ratio of B/G as summarized in Fig. 5.7.

This behavior is fundamentally related to the fact most common materials better resist a change in volume (high bulk modulus B compared

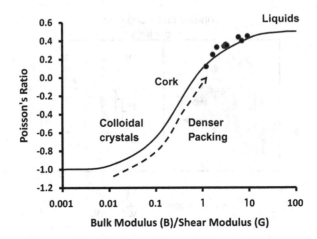

Fig. 5.7   Poisson's ratio as a function of the ratio of the bulk to the shear modulus. The solid curve describes the trend over a wide range of different material classes. [greaves11]

Table 5.2   A list of the bulk and shear modulus for some common materials of relevance to AFM. The value for Poisson's ratio are also provided.

Material	B - Bulk (GPa)	G - Shear (GPa)	B/G	Poisson ratio
Quartz, SiO_2	36	30	1.20	0.12
Glass	50	30	1.67	0.25
Steel	181	89	2.03	0.33
Copper	131	45	2.91	0.34
Aluminum	75	24	3.13	0.34
Indium	71	22.5	3.16	0.36
Polystyrene	4	1.2	3.33	0.35
Gold	166	28	5.93	0.44
Polyethylene	0.35	0.05	7.00	0.4
Lead	50	5.4	9.26	0.45

to G) than a change in shape by angular distortion (high shear modulus G compared to B). At the microscopic level, this implies that interatomic bonds prefer to align with a deformation rather than readjust by stretching or compressing. Thus when an AFM tip exerts a force, the value of ν for both the substrate and tip will play a role in the deformation that results.

A few materials from Table 5.2 are included as solid data points in Fig. 5.7 to better make this point.

From Fig. 5.7, it is evident that most samples studied by an AFM will have a Poisson's ratio that lies between values of 0.2 and 0.5. For most metals, ceramics and glasses, ν is about $1/3$ while for polymers and elastomers, ν tends to be closer to $1/2$.

If the Poisson's ratio is known for a homogeneous, isotropic material, we also have two simple equations relating E (Young's modulus) to the shear and bulk moduli G and B that are given by

$$E = 2G(1 + \nu), \tag{5.3a}$$

$$E = 3B(1 - 2\nu). \tag{5.3b}$$

If ν is approximated by $1/3$, these equations give the following estimates for G and B:

$$G = \frac{3E}{8} \quad \text{and} \quad B = E \quad \left(\text{when } \nu = \frac{1}{3}\right). \tag{5.4}$$

A systematic comparison of the ratio of G to E for polycrystalline metallic chemical elements has revealed an average value of 0.396, in close agreement with the expected value of $3/8$ in Eq. (5.4) [ledbetter77]. To make $E > 0$ in Eqs. (5.3), the resulting range of values for ν ($-1 < \nu < \frac{1}{2}$) can be viewed as a physical constraint, since a material will only be stable if the stiffness (elastic moduli) are positive.

It is also worth briefly mentioning that materials subjected to forces in more than one direction will require both E and ν or G and ν to specify the net deformation. Suppose a normal force F_x acts along the $+x$-direction in Fig. 5.5, and a second normal force F_z is now applied along the $+z$-direction. The resulting strain $\Delta L/L$ will contain two contributions, one due to the elongation in the x-direction $\sigma_x = F_x/(w \cdot t)$, and one due to the stress in the z-direction $\sigma_z = F_z/(L \cdot t)$. Assuming the block is elastically isotropic, σ_x will produce a strain in the x-direction equal to $+\sigma_x/E$ (see Eq. (5.1)). The stress σ_z will *decrease* the strain along x since elongation in z produces a *lateral contraction* along x (see Eq. (5.2)). The contribution to the net strain depends on Poisson's ratio and will be given by $-\nu\sigma_z/E$. The net strain in x (ε_x) will be the sum of the two opposing contributions: $\frac{\Delta L}{L} \equiv \varepsilon_x = \frac{1}{E}(\sigma_x - \nu\sigma_z)$. For a set of arbitrary forces in an elastically anisotropic material, tensor notation quickly becomes a necessity.

Example 5.1:   A cube of material 25 mm on a side has a Poisson ratio of 0.27. An outward force of 100 N applied to opposite faces of this material causes the length of the cube to elongate by 200 nm.

(a) What is the normal stress on the cube?

$$\text{stress} = \frac{F_n}{A} = \frac{100\,\text{N}}{(0.025\,\text{m})^2} = 1.6 \times 10^5\,\text{Pa} = 1.6 \text{ atmospheres}$$

(b) What is the normal strain in the cube?

$$\text{strain} = \frac{\Delta L}{L} = \frac{200\,\text{nm}}{(0.025\,\text{m})} = 8 \times 10^{-6}$$

(c) What is Young's modulus?

$$E = \frac{\text{stress}}{\text{strain}} = \frac{1.6 \times 10^5\,\text{Pa}}{8 \times 10^{-6}} = 20\,\text{GPa}$$

(d) Because of the extension of the cube along the direction of the applied force, what must be the change in the dimensions of the cube in the directions perpendicular to its extension? (See Fig. 5.5).

$$\nu = -\frac{\frac{\Delta w}{w}}{\frac{\Delta L}{L}} \Rightarrow \Delta w = -\nu \left(\frac{\Delta L}{L}\right) w$$

$$\Delta w = -0.27(8 \times 10^{-6})(0.025\,\text{m}) = -54\,\text{nm}.$$

At the microscopic level, the magnitude of the elastic modulus depends on two factors, the stiffness of the bond between individual atoms and the density of bonds per unit area. Chemical bonds can be modeled as springs and they can be characterized by a spring constant k (units of N/m). The number of bonds per unit area must be inversely proportional to the atomic size which can be modeled by a single parameter a_o (units of m). Thus you might expect that Young's modulus is roughly given by

$$E \simeq \frac{k}{a_o}. \tag{5.5}$$

This relationship is important because it establishes proportionality between an atomic spring constant and Young's modulus.

Since the spacing between atoms in most solids are all roughly the same ($a_o \cong 0.2 - 0.3\,\text{nm}$), the wide range of values for E is mostly related to the variation in k between different material classes. The weakest bond determines the measured modulus. Thus while k is about $\sim$500 N/m for a single covalent bond ($\sim$1000 N/m for double bond, $\sim$1500 N/m for triple bond), polymers contain both strong covalent bonds (in-chain) and weak inter-chain hydrogen bonds ($k = 0.5 - 2\,\text{N/m}$). Therefore, when a polymer is stressed, it is the weak hydrogen bonds between polymer chains that are first disrupted to give the large measured deformations and resulting low values for Young's modulus.

5.2 Contact Mechanics

Once a tip comes into contact with a substrate, it is useful to calculate the stresses and deformations which arise. In this section, we will review a few of the many models that are based on continuum mechanics already developed. Derivations of the various results will not be presented since they are already covered in sufficient detail in many textbooks on mechanical deformations. The models require estimates of the elastic modulus E and the Poison ratio ν for both the tip and substrate.

The tip shape is important in determining the deformation of the interface between the tip and sample. This point is made clearly by reviewing the results of a few classical models that predict the deformation caused by an applied force for different shaped tips. A few results are collected in Fig. 5.8 which shows the predictions for the deformation D of a substrate for three different cases. A result attributed to Boussinesq in 1885 considers the deformation of a substrate with modulus E and the Poison ratio ν when a delta-function like force F is applied at a point. As shown in Fig. 5.8, the resulting deformation $D(r)$ has a singularity as the radial distance from the point force approaches zero.

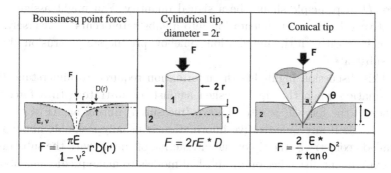

Fig. 5.8   The shape of an indenter determines the deformation D that results for an applied force F. Indenters with three different shapes are illustrated: a point force, a cylindrical tip of radius r, and a conical tip of half angle θ. Note that the point force produces a singularity as $r \to 0$.

A more realistic treatment of indentation models the tip as a solid cylinder with a diameter $2r$. Lastly, a final case is considered where a conical tip of half-angle $\gamma = \pi/2 - \theta$ is pushed into a substrate with a force. The maximum deformation D that results for an applied force F is also given in Fig. 5.8.

The deformation caused by the cylindrical and conical tip must depend on the elastic modulus of both the tip and substrate. Although you might expect the tip to be much harder than the substrate, any deformation in the tip and substrate must add and this suggests that the process can be treated as two springs in series. This implies an effective elastic modulus E^* must be defined as

$$\frac{1}{E^*} = \left(\frac{1}{E_1} + \frac{1}{E_2} \right), \tag{5.6}$$

and this effective modulus must be used to predict the maximum deformation D caused by an applied force F. More detailed considerations indicate the effective or reduced modulus defined in Eq. (5.6) must be modified to include the Poisson ratio of the two materials, so a more exact definition for E^* is given by

$$\frac{1}{E^*} = \left(\frac{1 - \nu_1^2}{E_1} + \frac{1 - \nu_2^2}{E_2} \right). \tag{5.7}$$

Yet another quantity is often defined as E_{tot}

$$\frac{1}{E_{\text{tot}}} = \frac{3}{4} \left(\frac{1 - \nu_1^2}{E_1} + \frac{1 - \nu_2^2}{E_2} \right) = \frac{3}{4} \frac{1}{E^*}. \tag{5.8}$$

Also, D in principle should be a signed quantity. You might assign D a negative value if a net indentation into a substrate occurs. Conversely, D might be positive if the indentation somehow produces a protrusion above the substrate.

This discussion highlights the information required to understand the deformation when the tip is pushed against a substrate with a force F. Although the force F is often known (measured) during an AFM experiment, little if any quantitative information is usually available about the detailed geometry of the tip or the interaction of the tip with the substrate. An experimentalist relies on established models to make predictions about relevant quantities. As in the above discussion, we usually assume the tip and substrate are comprised of different materials that have a Young's modulus, Poisson ratio, etc. designated by E_1, ν_1,.... and E_2, ν_2,.... respectively.

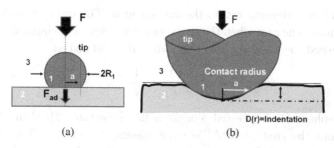

(a)   (b)

Fig. 5.9   In (a), a schematic of the deformation that occurs when a tip of material 1 is pushed against a flat substrate composed of material 2 by an external force F. The tip is approximated as a sphere of radius R_1. The tip and substrate are immersed in a surrounding medium designated as 3. Usually in AFM applications, 3 is either vacuum, air, or a liquid (water). The contact between the two materials can be characterized by the contact radius a. In (b), at the nanoscale, a deformation $D(r)$ also develops. The deformation is difficult to measure. Predicting the deformation usually relies on contact mechanic models.

To make progress, it is customary to approximate the tip as a sphere of radius R_{tip} interacting with a flat substrate as shown in Fig. 5.9. The sphere is pushed into the substrate with an applied force F. When the force F on the sphere is downward, the substrate exerts a repulsive force on the sphere, so the force is considered a positive number. This means an attractive force between the sphere and substrate is considered negative. For example, if the substrate exerts a downward force on the sphere, the force is considered negative. The geometry employed for the standard contact models is a sphere against a flat plane. In AFM, we interpret this sphere as the apex of the tip. It is important to remember in what follows that the spherical tip is not yet assumed to be attached to a flexible cantilever.

Quantities of interest include a, the contact radius of the tip with the substrate, F_{ad}, the adhesive force that develops between the tip and the substrate, the contact radius at zero load ($F = 0$), the contact radius at separation, and the pull-off force required to remove the tip from the substrate. Three classic theories — the Hertz model (1881), the Johnson–Kendall–Roberts (JKR) model (1971) and the Derjaguin-Muller-Troporov (DMT) model (1975) — developed to describe the contact between macroscale objects are usually invoked to predict these various quantities of interest.

5.3   Hertz Contact Mechanics

The classical Hertzian theory of contact focuses on the non-adhesive contact between two rigid, elastic spheres of radii R_1 and R_2 when negligible adhesion forces develop within the contact area. The theory therefore best approximates the case when the two spheres are rigid and elastically deform when pushed together. An effective radius R is defined as

$$\frac{1}{R} = \frac{1}{R_1} + \frac{1}{R_2} \qquad (5.9)$$

If a sphere (1) is pressed against a flat substrate (2), then $R_2 = \infty$ and within the context of AFM experiments, $R_1 = R_{tip}$. The Hertzian theory ignores any long range vdW forces that develops between the two objects. This model is reasonable when van der Waals, electrostatic, and chemical forces are negligible compared to the repulsive elastic interaction.

The fundamental predictions based on the Hertz model for the pull-off force (F_{ad}^{Hertz}), the contact radius (a_{Hertz}) and indentation (D_{Hertz} — here defined as a positive quantity) as a function of the applied force F are listed in Eqs. (5.10a–c).

$$F_{ad}^{Hertz} = 0, \qquad (5.10a)$$

$$a_{Hertz} = \left(\frac{R_{tip}F}{E_{tot}}\right)^{\frac{1}{3}}, \qquad (5.10b)$$

$$D_{Hertz} = \frac{a_{Hertz}^2}{R_{tip}}. \qquad (5.10c)$$

Example 5.2:   Assume a tip of radius 30 nm contacts a flat substrate. If the tip has an elastic modulus of 200 GPa and the substrate has a modulus of 100 GPa, calculate the contact radius, the deformation and the pull-off force for an applied force of 10 nN. Use the Hertz model. Assume both the tip and substrate has a Poisson's ratio of 0.3.

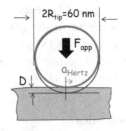

(Continued)

Example 5.2:   (*Continued*)

The effective modulus must first be calculated

$$\frac{1}{E_{tot}} = \frac{3}{4}\left(\frac{1-\nu_{sub}^2}{E_{sub}} + \frac{1-\nu_{tip}^2}{E_{tip}}\right)$$

$$= \frac{3}{4}\left(\frac{1-(0.3^2)}{100\,\text{GPa}} + \frac{1-(0.3)^2}{200\,\text{GPa}}\right)$$

$$= \frac{3(0.91)}{4}\left(\frac{1}{100\,\text{GPa}} + \frac{1}{200\,\text{GPa}}\right) = 0.01024\,\text{GPa}^{-1}$$

$$E_{tot} = 97.7\,\text{GPa}$$

(a) For the conditions given, the Hertz contact radius is obtained from Eq. (5.10b)

$$a_{Hertz} = \left(\frac{R_{tip}F}{E_{tot}}\right)^{\frac{1}{3}}$$

$$= \left(\frac{30 \times 10^{-9}\text{m} \cdot 10 \times 10^{-9}\text{N}}{97.7 \times 10^9 \text{Pa}}\right)^{\frac{1}{3}}$$

$$= 1.5 \times 10^{-9}m = 1.5\,\text{nm}$$

If we assume atoms are spaced by ~0.3 nm, a contact of this size will include about

$$N \approx \frac{\pi a_{Hertz}^2}{(a_o)^2} = \frac{\pi(1.5 \times 10^{-9}\,\text{m})^2}{(0.3 \times 10^{-9}\,\text{m})^2} \approx \text{about 78 atoms}$$

(b) The deformation D is given by Eq. (5.10c)

$$D_{Hertz} = \frac{a_{Hertz}^2}{R_{tip}}$$

$$= \left(\frac{F^2}{R_{tip}E_{tot}^2}\right)^{\frac{1}{3}}$$

$$= \left(\frac{(10 \times 10^{-9}\text{N})^2}{(30 \times 10^{-9}\,\text{m})(97.7 \times 10^9\,\text{Pa})^2}\right)^{\frac{1}{3}}$$

$$= 7.04 \times 10^{-11}\,\text{m} = 70.4\,\text{pm}$$

(c) The pull-off force, given by Eq. (5.10a) is zero.

(*Continued*)

Example 5.2: (*Continued*)

It is interesting to estimate the contact pressure that develops.

$$P \simeq \frac{F}{\pi a_{\text{Hertz}}^2} = \frac{10 \times 10^{-9} N}{\pi (1.5 \times 10^{-9}\, m)^2} \simeq 1.4 \times 10^9\, \text{Pa}.$$

This calculated pressure is large, ~14,000 times larger than atmospheric pressure. This estimate may be questioned because it (i) assumes the applied force is constant across the contact and (ii) uses a continuum theory for a contact that only contains ~78 atoms.

5.4 DMT Contact Mechanics

Like Hertz, the DMT contact model also describes the contact of an elastic sphere with adhesion against a rigid plane surface. DMT extends the Hertzian model because it allows for adhesion between the sphere and substrate after contact is made.

The DMT model predicts the adhesive pull-off force ($F_{\text{ad}}^{\text{DMT}}$), the contact radius ($a_{\text{DMT}}$) and indentation ($D_{\text{DMT}}$ — here defined as a positive quantity) as a function of the applied force F. The important results are listed in Eqs. (5.11a–c) [derjaguin75].

$$F_{\text{ad}}^{\text{DMT}} = 2\pi R_{\text{tip}} W_{132}, \tag{5.11a}$$

$$a_{\text{DMT}} = \left(\frac{R_{\text{tip}}[F + F_{\text{ad}}^{\text{DMT}}]}{E_{\text{tot}}} \right)^{\frac{1}{3}}, \tag{5.11b}$$

$$D_{\text{DMT}} = \frac{a_{\text{DMT}}^2}{R_{\text{tip}}}. \tag{5.11c}$$

The DMT model includes long range vdW forces between atoms in the sphere and the substrate outside the contact area. The model is valid for low adhesion, relatively stiff contacts that are made in air or vacuum under dry conditions. Even at low humidity, capillary condensation can occur between the tip and sample, a phenomenon which is not included in this model. DMT is most appropriate for hard substrates that experience low adhesion after contact and is widely used to understand AFM indentation data.

In the DMT model, the adhesive force $F_{\text{ad}}^{\text{DMT}}$ that develops when a tip of radius R_{tip} contacts a substrate (see Eq. 5.11(a)) is given by

$$F_{\text{ad}}^{\text{DMT}} = 2\pi R_{\text{tip}} W_{132}, \tag{5.12}$$

where W_{132} (J/m^2) is defined as the energy required to pull the two surfaces apart (work of adhesion). For clean surfaces, the energy expended in pulling-off is largely related to differences in surface energies and hence W_{132} is also known as surface energy adhesion. For this reason, any discussion of W_{132} is analogous to the discussion of surface energies already covered in Chapter 4.

In many circumstances, W_{132} is sufficiently small (less than $\sim 100\,\text{mJ/m}^2$) to produce any significant adhesion between a sphere and substrate. However within the context of AFM where nN forces are relevant, even small values for W_{132} can have important consequences. When tips and/or substrates are specially coated with monolayers of molecules to systematically measure W_{132}, the method is often referred to as chemical force microscopy. Estimates for W_{132} between a given spherical tip and substrate (consistent within the DMT model) can be obtained from Eq. (5.11a) by measuring the pull-off force $F_{\text{ad}}^{\text{DMT}}$.

When the applied force F is equal to zero, DMT predicts a finite radius of contact due to the long-range interaction between particles in the tip and substrate. From Eq. (5.11b), we find

$$a_{\text{DMT}}(F = 0) = \left(\frac{2\pi W_{132} R_{\text{tip}}^2}{E_{\text{tot}}} \right)^{\frac{1}{3}}. \tag{5.13}$$

Recall that when separating a spherical tip (1) from a substrate (2), if the tip and substrate are comprised of different materials, the energy required to break the existing tip–substrate interface is always less than the surface energy of the two new surfaces created. The reduction in energy is attributed to the interfacial energy γ_{12} which describes how the atoms in material 1 interact with the atoms in material 2 when they are in close contact.

The work required breaking an existing interface between a tip and substrate in the presence of a third medium (3) is formally given by the Dupre equation

$$W_{132} = \gamma_{13} + \gamma_{23} - \gamma_{12}. \tag{5.14}$$

For weakly interacting materials, the values of γ used in Eq. (5.14) will be dominated by vdWs forces as discussed in Chapter 4. If the materials 1 and 2 strongly interact, they may form chemical bonds and the value of W_{132} must reflect this increased interaction. In this case, W_{132} quantifies the contribution of many parallel bonds that are often modeled as a statistical distribution of bonds in various states of disruption.

According to Eqs. (5.11(a–c)), the adhesive force $F_{\text{ad}}^{\text{DMT}}$ must be added to the applied force F to predict both the contact radius a_{DMT} and the deformation D_{DMT}. The new feature in DMT is the existence of a contact radius and finite deformation even when the applied force between the tip and substrate is zero. A comparison between the predictions of Hertz and DMT for the deformation D as a function of the applied force F is given in Fig. 5.10.

The deformation D plotted in Fig. 5.10 is the combined deformation of the tip and substrate for a system with $E_{\text{tot}} = 97.7$ GPa and $R_{\text{tip}} = 30$ nm. There are many ways to achieve this condition. For instance if the tip is infinitely stiff ($E_1 \to \infty$), then $E_{\text{tot}} = 97.7$ GPa if, for example, $\nu_2 = 0.3$ and the substrate modulus is given by $E_2 = 66.6$ GPa. In this situation, the entire deformation D would occur in the substrate. Alternatively, as shown in Example 5.2, the condition $E_{\text{tot}} = 97.7$ GPa also pertains if $\nu_1 = \nu_2 = 0.3$ and $E_1 = 200$ GPa, $E_2 = 100$ GPa. In this case, both the tip and substrate will deform with the total deformation for a given applied force as plotted in Fig. 5.10.

Also shown in Fig. 5.10 is the influence of W_{132} on the deformation. As $W_{132} \to 0$, DMT $\to$ Hertz.

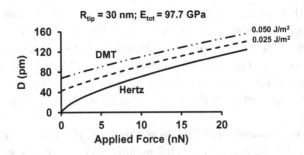

Fig. 5.10   A comparison between the predictions of the Hertzian and DMT models for deformation (in picometers) as a function of applied force (in nanoNewtons). The same parameters are used for E_1, E_2, R_{tip}, etc. as given in Example 5.2. Two representative values for W_{132} illustrate the effect of this parameter on the DMT model. Note that DMT predicts a finite deformation at zero applied force. Here, the deformation D is plotted as a positive quantity.

5.5 JKR Contact Mechanics

In contrast to the Hertz and DMT models for contact, the Johnson–Kendall–Roberts (JKR) model predicts a hysteretic adhesion due to the necking of the sample [johnson71]. The model includes only interaction forces inside the contact area, so there are no long range forces between the sphere and substrate as the sphere approaches the substrate. The basis for the model lies in a calculation of the total energy of the sphere-substrate system which includes (i) the stored elastic energy, (ii) the mechanical energy imparted by the applied load, and (iii) the surface energy. An equilibrium condition is obtained when the derivative of this energy with respect to the contact radius equals zero. This condition gives the results for the JKR model for the pull-off force ($F_{\text{ad}}^{\text{JKR}}$), the contact radius ($a_{\text{JKR}}$) and indentation ($D_{\text{JKR}}$) as a function of the applied force F as given in Eqs. (5.15a–c).

$$F_{\text{ad}}^{\text{JKR}} = \frac{3\pi}{2} R_{\text{tip}} W_{132}, \tag{5.15a}$$

$$a_{\text{JKR}} = \left(\frac{R_{\text{tip}}}{E_{\text{tot}}} \left[\sqrt{F_{\text{ad}}^{\text{JKR}}} + \sqrt{F + F_{\text{ad}}^{\text{JKR}}} \right]^2 \right)^{\frac{1}{3}}, \tag{5.15b}$$

$$D_{\text{JKR}} = \frac{a_{\text{JKR}}^2}{R_{\text{tip}}} - \frac{4}{3} \sqrt{\frac{F_{\text{ad}}^{\text{JKR}} a_{\text{JKR}}}{R_{\text{tip}} E_{\text{tot}}}}. \tag{5.15c}$$

The JKR contact model is more complicated than DMT because it is (i) non-conservative and (ii) requires a prior knowledge of the history of the tip-sample contact.

Within the JKR model, the adhesive force $F_{\text{ad}}^{\text{DMT}}$ that develops when a sphere of radius R_{tip} contacts a substrate is specified by

$$F_{\text{ad}}^{\text{JKR}} = \frac{3\pi}{2} R_{\text{tip}} W_{132}, \tag{5.16}$$

where W_{132} is defined by Eq. (5.14).

Any adhesive force required to describe highly deformable surfaces is better approximated by JKR than DMT. In JKR, the adhesive force $F_{\text{ad}}^{\text{JKR}}$ must be added to the applied force F to predict the contact radius a_{JKR} and deformation D_{JKR} as described in Eqs. (5.15a–c). Estimates for W_{132} between a given tip and substrate (consistent within the JKR model) can be obtained from Eq. (5.15a) by measuring the maximum pull-off force required to overcome $F_{\text{ad}}^{\text{JKR}}$. Such a measurement is complicated because the contact deforms so that the measured pull-off force is usually less than the maximum pull-off force.

When the applied force $F = 0$ in Eq. (5.14b), JKR predicts a finite radius of contact given by

$$a_{\text{JKR}}(F = 0) = \left(\frac{6\pi W_{132} R_{\text{tip}}^2}{E_{\text{tot}}} \right)^{\frac{1}{3}}. \tag{5.17}$$

A new feature in the JKR model occurs when the spherical tip is retracted from the substrate. The tip–sample contact persists until a critical radius is reached at which point the two surfaces abruptly separate. The contact radius at separation can be found when the applied force just cancels the adhesive force. The contact radius at separation ($a_{\text{JKR}}^{\text{separation}}$) is then found by setting $F + F_{\text{ad}}^{\text{JKR}} = 0$ in Eq. (5.15b)

$$a_{\text{JKR}}^{\text{separation}} = \left(\frac{F_{\text{JKR}}^{\text{ad}} R_{\text{tip}}}{E_{\text{tot}}} \right)^{\frac{1}{3}} = \left(\frac{3\pi W_{132} R_{\text{tip}}^2}{2E_{\text{tot}}} \right)^{\frac{1}{3}}. \tag{5.18}$$

Because the JKR model is most applicable to soft, compliant samples, this transition is preceded by the formation of a material bridge which develops between the tip and substrate.

Note that the representation of the JKR model does not include any long range attractive interaction forces when the tip approaches the substrate. Upon sphere–substrate contact, a contact radius forms and an adhesive force develops from the short range interaction between particles in the tip and substrate. This adhesive force pulls the tip toward the substrate and produces a deformation even under zero applied force. As the applied force is increased, the deformation and the contact radius increase. When the applied force is decreased and returned to zero, the non-conservative nature of the JKR interaction causes a different behavior. The tip–substrate contact radius decreases during tip withdrawal until the critical contact radius is reached when the tip suddenly breaks free from the substrate.

The deformation D predicted by the JKR model as a function of the applied force F is given in Fig. 5.11. The plot is now more complicated and must include both positive and negative values for the deformation due to the complexity of the JKR model. In Fig. 5.11, a compliant substrate ($E_2 = 0.1\,\text{GPa}$) is indented by a spherical tip ($E_1 = 10\,\text{GPa}$) with $R_{\text{tip}} = 40\,\text{nm}$. Values assigned to the other relevant parameters are: $\nu_1 = \nu_2 = 0.3$ and $W_{123} = 0.05\,\text{J/m}^2$. In this case, $E_{\text{tot}} = 0.145\,\text{GPa}$. The deformation D plotted is the combined deformation of the spherical tip and substrate, but because the spherical tip is stiffer than the substrate, the deformation is essentially that of the substrate.

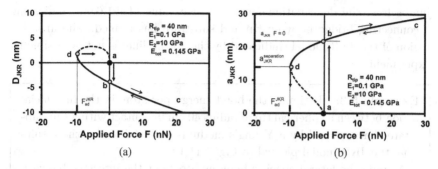

(a)　　　　　　　　　　　　　　(b)

Fig. 5.11   In (a), the deformation predicted by the JKR model as a function of the applied force. The symbols a, b, c, d, are described in the text. The spherical tip has a modulus of $E_2 = 10.0\,\text{GPa}$ and is pushed into a soft substrate with a modulus of $E_1 = 0.1\,\text{GPa}$. The tip radius is taken to be $40\,\text{nm}$. Values assigned to the other relevant parameters are: $\nu_1 = \nu_2 = 0.3$ and $W_{123} = 0.05\,\text{J/m}^2$. The deformation is now a signed quantity to account for the indentation $(-)$ and elongation $(+)$ predicted by the JKR model. In (b), the variation of the contact radius as a function of the applied force. The dotted lines indicate unstable regions that are not accessible by experiment.

Initially, at point a, the spherical tip is resting on the substrate. When the JKR adhesion is turned on, the deformation of the sphere abruptly increases as indicated by the vertical line $a \to b$ in Fig. 5.11(a) and a deformation at zero applied force develops. Here the deformation is a negative quantity. At the same time, a finite contact radius of $\sim 21.9\,\text{nm}$ (see Eq. (5.17)) also forms as illustrated in Fig. 5.11(b). As additional force is applied, both the deformation and contact radius increase along the line $b \to c$. At point c, the applied force is caused to decrease and the deformation and contact radius retrace along $c \to b \to d$.

During withdrawal of the spherical tip, at point b, the applied force is equal to zero. As the applied force becomes negative (indicative of a pulling force), the deformation switches from negative to positive. This indicates the onset of a material bridge that forms between the substrate and sphere due to the adhesion force specified in Eq. (5.16).

When going from $b \to d$, the sphere pulls away from the substrate and the deformation eventually becomes positive, indicating a material bridge has formed between the tip and substrate. Meanwhile, the contact radius continues to shrink. During retraction, the tip finally exerts a force that is equal and opposite to the adhesive force of $9.4\,\text{nN}$ predicted by Eq. (5.16). Under this condition, the contact radius reaches the critical value given by Eq. (5.18) ($13.8\,\text{nm}$ in Fig. 5.11(b)) and the spherical tip snaps free from the substrate at point d. When the spherical tip jumps free, the deformation has a positive value of +1.6 nm, indicating a connecting bridge between tip and substrate has formed. The unstable region of the JKR model (indicated by the dotted line) is not accessible by experiment.

Example 5.3:   Estimate the bond energy per atom in the interfacial region between a spherical tip of radius 40 nm in contact with a soft substrate characterized by a Young's modulus of 0.1 GPa. Use the results from the JKR model plotted in Fig. 5.11(b).

Assume the lateral spacing between atoms at the interface between the spherical tip and substrate is $a_o \approx 0.3$ nm. At separation, Fig. 5.11(b) gives $a_{\text{JKR}}^{\text{separation}} \simeq 13.8$ nm. The approximate number of atoms in the contact just before breaking is roughly

$$N \approx \frac{(a_{\text{JKR}}^{\text{separation}})^2}{(a_o^2)} = \frac{(13.8)^2}{(0.3)^2} = 2120.$$

The area just before breaking is

$$A \approx \pi(a_{\text{JKR}}^{\text{separation}})^2 = 600 \, \text{nm}^2 = 6 \times 10^{-16} \, \text{m}^2.$$

The energy released when the contact is broken can be estimated as

$$E_{\text{bonds}} \approx W_{132}A = (0.05 \, J/m^2)(6 \times 10^{-16} \, m^2)$$

$$= 3 \times 10^{-17} J \cdot \left[\frac{1 \, eV}{1.6 \times 10^{-19} J} \right] = 190 \, eV.$$

The energy per bond is therefore

$$\Delta U \approx \frac{E_{\text{bonds}}}{N} = \frac{190 \, \text{eV}}{2120} = 0.09 \, \text{eV} = 90 \, \text{meV/bond}.$$

5.6   Maugis-Dugdale Theory

The contact theories described above (Hertz, DMT, JKR) were originally thought to be distinct, but work by Maugis[maugis92] showed they were interconnected by a dimensionless elasticity parameter λ that measured the ratio of adhesion to elasticity. In brief, Maugis used the Dugdale approximation [dugdale60] that the adhesive (attractive, tensile) force remains constant until a separation distance specified by h is reached. The predictions are known as the Maugis–Dugdale (M–D) theory.

Formerly, λ is defined as

$$\lambda = \frac{2}{a_0}\left(\frac{9R_{\text{tip}}W_{132}^2}{16\,\pi E_{\text{tot}}^2}\right)^{\frac{1}{3}} = 1.65\frac{1}{a_0}\left(\frac{R_{\text{tip}}W_{132}^2}{\pi E_{\text{tot}}^2}\right)^{\frac{1}{3}} \quad \text{[no dimensions]},$$

(5.19)

where E_{tot} is defined in Eq. (5.8) and a_o is a parameter that specifies the characteristic spacing between atoms (~ 0.2 to $\sim 0.3\,\text{nm}$). A dimensionless force parameter $\bar{F}$ is also defined

$$\bar{F} = \left(\frac{F}{\pi W_{132}R_{\text{tip}}}\right) \quad \text{[no dimensions]}.$$

(5.20)

The range of λ where each contact mechanics model is valid is indicted in Fig. 5.12. When λ is small (typically less than 0.1), the DMT model applies. As λ increases, a transition region is encountered (labelled M–D) before the JKR model is generally applicable for $\lambda > 6$.

In the Maugis transition regime, a Dugdale potential is used to model the separation energy of the contact. In contrast to a 12-6 (Lennard-Jones) potential, the Dugdale potential is a piece-wise linear approximation which relies on an adjustable length parameter h.

When the surface-to-surface separation d is less than h ($d < h$), the Dugdale interaction potential energy has a constant slope. When $d > h$, the interaction potential energy is zero. The corresponding force is therefore constant for distances $d < h$ and is zero for separations $d > h$.

The limits of validity for the M–D model are indicated by the shaded region in Fig. 5.12. M–D assumes a Dugdale interaction within an annular zone that surrounds the contact area. The extent of this zone is

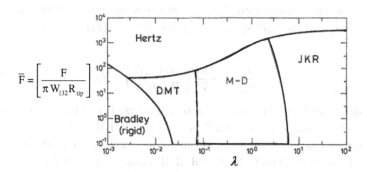

Fig. 5.12 The appropriate contact model can be determined by calculating the reduced force $\bar{F}$ and the dimensionless elasticity parameter λ defined in Eqs. (5.19)–(5.20). [johnson97].

characterized by a dimensionless parameter m which measures the ratio of the width of the annular region to the radius of the contact area. Predictions for deformation using M–D are not analytical as in Hertz, DMT, JKR and some computational effort must be expended. Below we lay out the ingredients of a calculation and include only the relevant formulas.

Operationally, the modulus of the substrate and tip are first used to estimate an effective modulus, E_{tot}. If the tip radius R_{tip} and work of adhesion are specified, then λ (see Eq. (5.19)) can be estimated. Knowing λ, the resulting contact radius a is determined as a function of a parameter m (typically $100 < m < 1$) by solving the parametric Eq. (5.21) which is quadratic in a

$$1 = \frac{\lambda a^2}{2} \left(\frac{E_{tot}}{\pi R_{tip}^2 W_{132}} \right)^{\frac{2}{3}} \left[\sqrt{m^2 - 1} + (m^2 - 2)\arctan\left(\sqrt{m^2 - 1} \right) \right]$$

$$+ \frac{4\lambda^2 a}{3} \left(\frac{E_{tot}}{\pi R_{tip}^2 W_{132}} \right)^{\frac{2}{3}} \left[1 - m + \sqrt{m^2 - 1}\arctan\left(\sqrt{m^2 - 1} \right) \right].$$

$$(5.21)$$

For a given m, the positive solution for a can then be used to determine the force F required to produce a using

$$F = \frac{E_{tot} a^3}{R_{tip}} - \lambda a^2 \left(\frac{\pi W_{132} E_{tot}^2}{R_{tip}} \right)^{\frac{2}{3}} \left[\sqrt{m^2 - 1} + m^2 \arctan\left(\sqrt{m^2 - 1} \right) \right].$$

$$(5.22)$$

Finally, the deformation D_{M-D} consistent with the contact radius a (Eq. (5.21)) and the corresponding applied force F (Eq. (5.22)) is obtained from

$$D_{M-D} = \frac{a^2}{R_{tip}} - \frac{4\lambda a}{3} \left(\frac{\pi W_{132}}{R_{tip} E_{tot}} \right)^{\frac{1}{3}} \sqrt{m^2 - 1}. \qquad (5.23)$$

By systematically varying m, a plot of D_{M-D} vs. F can be obtained which bridges the regime between DMT ($\lambda < 0.1$) and JKR ($\lambda > 6$).

A representative comparison of the deformation as a function of applied force for the Hertz, DMT, M–D and JKR models is provided in Fig. 5.13.

Such a plot can be somewhat misleading since only *one* of the models will be valid for the set of parameters chosen. The specific parameters used to generate this plot are specified in the figure caption and give a

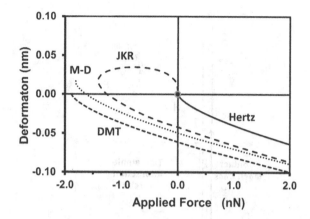

Fig. 5.13 A composite plot showing the deformation for different contact mechanic models. The parameters used are $E_1 = 70\,\mathrm{GPa}$, $E_2 = 20\,\mathrm{GPa}$, $W_{132} = 0.01\,\mathrm{J/m^2}$, $R_{\mathrm{tip}} = 30\,\mathrm{nm}$, and $a_o = 0.3\,\mathrm{nm}$. These parameters give $\lambda = 0.08$, a value that falls near the boundary between DMT and M–D.

value of $\lambda = 0.08$, indicating a situation that was intentionally selected to border between the DMT and M–D regimes. Figure 5.13 indicates the M–D model predicts a small material bridge (positive D) between the tip and substrate whereas the DMT predicts no such bridge. The finite deformation when $F = 0$ is a clear indicator of the influence of the adhesive forces.

By way of summary, using these four contact models quantitative calculations can be performed to assess the deformation caused by a tip exerting a force in an AFM experiment. Realistic estimates of the elastic properties of both the tip and substrate material are required. In addition, knowledge of the tip radius is also required.

5.7 The Tip–Sample Interaction

By combining the results in Chapters 4 and 5, we are finally in a position to discuss realistic models for a spherical tip–flat surface interaction. There are two regimes as indicated schematically in Fig. 5.14: (i) the long-range tip–sample interaction force that gives way to (ii) contact mechanics when the spherical tip is separated by a distance of a_o from the sample. As illustrated in Fig. 5.15, various models can be used to characterize this interaction depending on the level of sophistication that is desired.

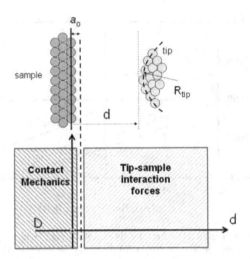

Fig. 5.14   Defining the two important regimes probed by an AFM as the tip approaches a flat substrate. As the tip approaches the substrate, vdW forces dominate the tip-substrate interaction. When the tip makes contact with the sample, contact mechanics must be used. The typical spacing between atoms a_o defines the transition region.

Perhaps the simplest model to consider is an infinitely hard tip and a flat sample with negligible long-range interaction forces. The situation is shown schematically in Fig. 5.15(a) which indicates the absence of long range forces as the tip–sample distance d is decreased. When $d = a_o$, the tip is considered in-contact with the sample and contact mechanic deformations can then occur. Because both the tip and sample are considered infinitely stiff, no indentation results and the interaction force rapidly increases with no deformation as shown.

A more realistic model incorporates Hertz contact mechanics as shown in Fig. 5.15(b). This model is appropriate when the tip and sample are hard and exhibit no adhesion. Here, the deformation D relative to a_o after contact is plotted as a negative number.

Finally, if long-range interaction forces are important as well as adhesion, and the sample and tip are reasonably hard, the DMT model provides a good approximation to the situation. It is customary to model the interaction force as piecewise continuous using results from Eqs. (4.10), (5.11) and (5.8)

$$
F_{\text{DMT}}(d) = \begin{cases} -\dfrac{HR_{\text{tip}}}{6d^2} & d > a_o \\[2ex] -\dfrac{HR_{\text{tip}}}{6a_o^2} + \dfrac{4}{3}E^*\sqrt{R_{\text{tip}}}(a_o - D)^{\frac{3}{2}} & D \leq a_o \end{cases} , \quad (5.24)
$$

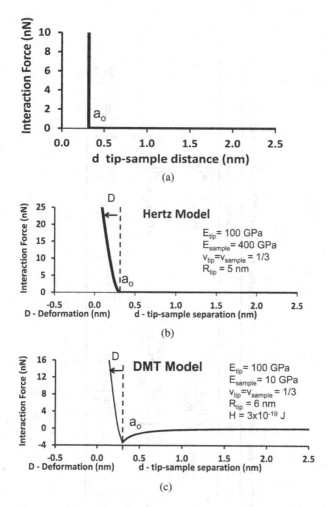

Fig. 5.15 Three models of the tip–substrate interaction assuming the tip is spherical. The various parameters used are listed in each graph. The x-axis has a dual function: when $d > a_o$, the x-axis measures the tip-sample surface-to-surface separation. When $d < a_o$, the x-axis is used to track the deformation D as a function of the applied force.

where

$$E^* = \left[\frac{1 - \nu_{\text{tip}}^2}{E_{\text{tip}}} + \frac{1 - \nu_{\text{sample}}^2}{E_{\text{sample}}} \right]^{-1}.$$

The above discussion assumes a tip that is clean and a sample that is flat. More sophisticated models are required when the tip contacts a very soft sample, when the tip encounters a thin film on a substrate, when the

Table 5.3 A collection of relevant results predicted by the Hertz, DMT and JKR contact models.

	$F_{\text{pull-off}} = F_{\text{ad}}$	Contact Radius	Deformation
Hertz	0	$a_{\text{Hertz}} = \left(\dfrac{R_{\text{tip}}F}{E_{\text{tot}}}\right)^{\frac{1}{3}}$	$D_{\text{Hertz}} = \dfrac{a_{\text{Hertz}}^2}{R_{\text{tip}}}$
DMT	$F_{\text{ad}}^{\text{DMT}} = 2\pi R_{\text{tip}} W_{132}$	$a_{\text{DMT}} = \left(\dfrac{R_{\text{tip}}[F + F_{\text{ad}}^{\text{DMT}}]}{E_{\text{tot}}}\right)^{\frac{1}{3}}$	$D_{\text{DMT}} = \dfrac{a_{\text{DMT}}^2}{R_{\text{tip}}}$
JKR	$F_{\text{ad}}^{\text{JKR}} = \dfrac{3\pi}{2} R_{\text{tip}} W_{132}$	$a_{\text{DMT}} = \left(\dfrac{R_{\text{tip}}}{E_{\text{tot}}}\left[\sqrt{F_{\text{ad}}^{\text{JKR}}} + \sqrt{F + F_{\text{ad}}^{\text{JKR}}}\right]^2\right)^{\frac{1}{3}}$	$D_{\text{JKR}} = \dfrac{a_{\text{JKR}}^2}{R_{\text{tip}}} - \dfrac{4}{3}\sqrt{\dfrac{F_{\text{ad}}^{\text{JKR}}}{R_{\text{tip}}}\dfrac{a_{\text{JKR}}}{E_{\text{tot}}}}$

sample is rough or when a water bridge forms between the tip and sample due to water condensation from the ambient air.

5.8 Chapter Summary

When an AFM tip contacts a substrate, significant forces can be generated that cause deformations to result. Following a general review of how solids deform when subjected to external forces, the elastic moduli relevant to AFM experiments were defined. Realistic estimates of these moduli are required to quantitatively interpret AFM experiments. For this reason, various plots and tables were provided to allow estimates for typical values of these moduli across a wide range of different materials.

When the tip contacts the substrate, contact mechanics are required to understand the deformations that occur. The predictions of three well-known models were discussed. The important predictions for the lift-off force at zero load, the contact radius a, and the deformation D for the Hertz, DMT and JKR contact models are collected in Table 5.3.

5.9 Further Reading

Chapter Five References:

[ashby89] M.F. Ashby, C Acta Metal. **37**, 1273–93 (1989).

[cardarelli08] F. Cardarelli, *Materials Handbook: A Concise Desktop Reference*, 2nd edition. Springer–Verlag, London (2008).

[derjaguin75] B.V. Derjaguin, V.M. Muller, Y.P. Toporov, "Effect of contact deformation on the adhesion of particles", J. Colloid and Interface Science **53**, 314–320 (1975).

[dugdale60] D.S. Dugdale, Yielding od Steel Sheets Containing Slits, J. Mech. Phys. Solids **8**, 100–04 (1960).

[gorecki80] T. Gorecki, "Relations between the Shear Modulus, the Bulk Modulus and Young's Modulus for Polycrystalline Metallic Elements", Mat. Sci. Engin. **43**, 225 (1980).

[greaves11] G.N. Greaves, et al. Nature Materials **10**, 823 (2011).

[johnson71] K.L. Johnson, K. Kendall, A.D. Roberts, "Surface energy and the contact of elastic solids", Proc. R. Soc. Lond. A. **324**, 301–313 (1971).

[johnson97] K.L. Johnson and J.A. Greenwood, "An adhesion map for the contact of elastic spheres", J. of Colloid Interface Sci. **192**, 326–333, 1997.

[maugis92]   D. Maugis, "Adhesion of spheres: the JKR-DMT transition using a Dugdale model", J. Colloid Interface Sci. **150**, 243–69 (1992).
[ledbetter77] H.M. Ledbetter, "Ratio of Shear and Young's Moduli for Polycrystalline Metallic Elements", Mat. Sci. and Engin. **27**, 133 (1977).

Further reading:

Physical Constants, 9th Ed. by W.H.J. Childs (Halsted Press, a division of John Wiley and Sons, New York, 1972).
Physics and Chemistry of Interfaces by H.-J. Butt, K. Graf, M. Kappl, Wiley-VCH Verlag & Co. KGaA, Weinheim, Germany (2003).
Solid State Physics by Gerald Burns, Academic Press, New York (1985).

5.10   Problems

1. What are the units of the moduli of elasticity (E, G, and B) and Poisson's ratio ν in the metric system?

2. The theory of isotropic linear elasticity allows Poisson's ratio to lie in the range from -1 to $+1/2$. What physical property gives rise to a positive value for Poisson's ratio?

3. Two rods are made of the same material and have the same volume. Rod 1 has cross-sectional area A and rod 2 has cross-sectional area 3A. Experiment shows that the length of rod 1 increases by Δx when subjected to a normal force F. What normal force F' is needed to stretch rod 2 by the same amount?

4. One way to measure the shear modulus of a material is illustrated in the diagram below.

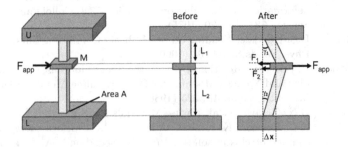

Two bars of a material with cross-sectional area $A = 10\,\text{mm} \times 10\,\text{mm}$ and lengths $L_1 = 75\,\text{mm}$ and $L_2 = 150\,\text{mm}$ are supported as shown. The upper support block U and lower support block L are held rigid while the middle block M rigidly clamps the two rods. Block M is subjected to an external applied force $F_{app} = 20\,\text{N}$. After the force is applied, careful measurements show the bars attached to block M are displaced by $\Delta x = 0.15\,\text{mm}$ as shown in the diagram. In equilibrium, two tangential forces F_1 and F_2 must develop parallel to the face of the two bars to counterbalance the applied force F_{app}.

(a) From this information, calculate the angles γ_1 and γ_2.
(b) Using the definition for the shear modulus, relate the angles γ_1 and γ_2 to the tangential forces F_1 and F_2 that must develop.
(c) Set the sum of F_1 and F_2 to equal F_{app} and solve for G.

5. Fused quartz or fused silica is glass consisting of silica in non-crystalline form. It differs from traditional glasses because it contains no additives which lower the melting temperature of glass. A straight line is scratched into a block of clear fused quartz. What shear strain is required to rotate this line by 0.5° (0.0087 radians)? Young's modulus of fused quartz is 72 GPa and Poisson's ratio is 0.17.

6. A rectangular bar of length $L = 100\,\text{mm}$ with a cross section of $10\,\text{mm} \times 10\,\text{mm}$ is subjected to a force F of 1,500 N. While under tension, the dimensions of the bar (L', w', t') are carefully measured as indicated in the diagram below. What is Young's modulus and Poisson's ratio for the material comprising the bar?

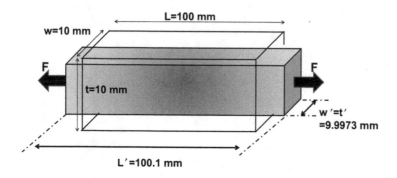

7. A square nanopillar of material is fabricated so it is perpendicular to a flat substrate. The pillar has a cross-section of 5 nm×5 nm and is 40 nm long. The pillar is fabricated from a material with a Young's modulus of 80 GPa and a Poisson's ratio of 0.3. How much does the length of the pillar compress when an AFM tip applies an external force of 10 nN

to the top of the pillar? What is the fractional change in the pillar's volume?

8. Consider a cube of dimension $L \times L \times L$. Assume the cube is elastically isotropic and has an elastic modulus E and a Poisson's ratio ν. If the cube is subjected to a hydrostatic pressure P, calculate that the resulting strain at right angles to any cube face. You should have a formula containing three contributions due to the strain that develops in the x, y and z directions. These three terms sum to give a net strain equal to

$$\varepsilon_{net} = -\frac{P}{E} + \frac{\nu P}{E} + \frac{\nu P}{E} = \frac{(2\nu - 1)}{E} P.$$

9. Using the Hertz contact model, what are the contact radius, the deformation and the pull-off force when a spherical tip (radius 50 nm, elastic modulus of 200 GPa) indents a flat substrate having a modulus of 75 GPa. Assume both the tip and substrate have a Poisson's ratio of 0.25 and that the tip is loaded against the substrate with an external force of 50 nN.

10. Using the DMT contact model, what are the contact radius, the deformation and the pull-off force when a spherical tip (radius 50 nm, elastic modulus of 200 GPa) indents a flat substrate having a modulus of 75 GPa. Assume both the tip and substrate have a Poisson's ratio of 0.25 and that the tip is loaded against the substrate with an external force of 50 nN. Assume the work of adhesion between the tip and substrate is 50 mJ/m^2.

Chapter 6

Quasi-Static Cantilever Mechanics

In the previous chapter, the force acting between objects that are in contact was discussed. The contact force when added to the non-contact forces discussed in Chapter 4 provides a model for the tip–substrate interaction in AFM. The magnitude of the force is small, typically less than 10 nN. How can these small forces be accurately measured? A force transducer must be used that converts the forces of interest into a measureable voltage signal. The force transducer used in the original implementation of an AFM in 1986 was the deflection of a microcantilever and the AFMs in use today still use this technique. The following chapter provides a discussion of the basic principles and assumptions that underlie this choice.

In what follows, we rely on standard principles developed to describe the mechanical deformation of solids. The focus of this chapter is to use these principles to interpret the deformations of a thin cantilever commonly used with an atomic force microscope. There are many systematic and elegant derivations of these results, but understanding rather than mathematical elegance is our goal. The discussion assumes a rectangular cantilever with one end fixed. The length of the cantilever is assumed to be much greater than its thickness. The math required to describe the thermal fluctuations of a thin cantilever is also considered. A discussion about the lateral deformation of a cantilever due to frictional forces is delayed until a further chapter.

The topics covered in this chapter are most accessible to students in mechanical engineering, materials science (with an emphasis on material deformation) and physics, since undergraduates in these disciplines tend to study the relevant concepts beyond the introductory course level.

6.1 The Deflection Spring Constant of a Thin Rectangular AFM Cantilever

The relevant geometry of the cantilever is shown in Fig. 6.1. Typically, a thin rectangular beam of length L, width w_c and thickness t_c is attached to a base support as shown. Even though a sharp tip is formed at the end of the beam as schematically shown, the effects of this tip on the mechanical deformation of the cantilever are not considered in the discussion that follows. A constant upward force F applied to the tip causes the end of the cantilever beam to deflect through a distance q. As a result, an unspecified shape, denoted by $w(x, t)$ will describe the resulting profile of the cantilever. Knowledge of the profile $w(x, t)$ for all x is a worthwhile goal. In general, the shape of the cantilever profile could depend on time t as well as the distance x, but for now we do not consider any time dependence. Hence the discussion that follows is often referred to as the **static cantilever model**.

The end of the cantilever $(x = L)$ deflects through a distance q and the deflected cantilever assumes an angle θ with respect to the neutral axis. Typical values for a rectangular microcantilever used in AFM are $L = 200\mu m$, $w_c = 30\mu m$ and $t_c = 2\mu m$. An extension to triangular cantilevers will be given in a separate chapter in Part II of the lecture notes.

At issue is to predict the effect of an applied force F (in the nN range) on the resulting deflection (either q or θ). For small deflections, the deflection is proportional to the applied force. Knowing the proportionality constant implies that by measuring either q or θ, the force can be inferred.

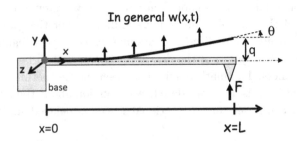

Fig. 6.1   A cantilever of length L, thickness is t_c and width w_c. The deflection from equilibrium is specified by the function $w(x, t)$. Upon application of an applied force F, the end of the cantilever deflects through a distance q and makes an angle θ with respect to the axis.

It is worthwhile to discuss in a heuristic way the salient points under-lying cantilever mechanics. Upon application of a point force $\vec{F}$ to the end of the cantilever, a torque will develop at a distance $\vec{r}$ from the origin that is specified by

$$\vec{\tau} = \vec{r} \times \vec{F} = |\vec{r}||\vec{F}| \sin \theta(r, F). \qquad (6.1)$$

For small deflections, $\vec{r} \perp \vec{F}$, and

$$|\vec{\tau}| = |\vec{r}||\vec{F}|. \qquad (6.2)$$

When torque refers to the bending of an object, it is often called a moment M rather than a torque τ. This word choice then makes it clear we are discussing elastic bending rather than a net rotation. For the case of the cantilever, the position vector $\vec{r}$ can lie anywhere along the cantilever beam, so we can define $|\vec{r}| = L - x$ where $0 \leq x \leq L$ and x is measured from the cantilever base as shown in Fig. 6.1. The magnitude of the bending moment M in $+\hat{z}$ will be given by

$$|\vec{M}| = (L - x)F. \qquad (6.3)$$

In equilibrium, all **internal** moments at x must balance the moment M developed by the applied force F at x. These internal moments are generated by the forces between atoms that comprise the cantilever. It is useful to imagine each atom at the location x is connected by a spring as shown schematically in the diagram sketched in Fig. 6.2. Each internal "spring" connecting adjacent atoms at x must then apply a moment around the neutral axis of the cantilever that tends to restore the cantilever to its original position.

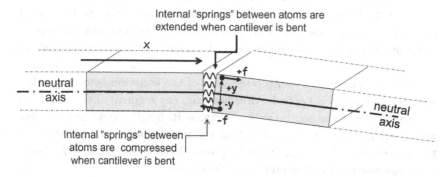

Fig. 6.2   An exploded view of the cantilever at a position x schematically showing the extension (for $+y$) and compression (for $-y$) of hypothetical "internal" springs between atoms. Along the neutral axis, there is no extension or compression.

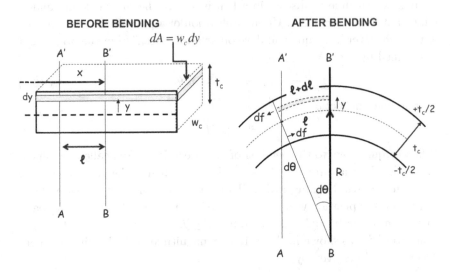

Fig. 6.3   An enlarged view of a cantilever segment of length dx located a distance x from the fixed end illustrating the effect of bending (greatly exaggerated). A thin section of the cantilever with thickness dy is located a distance $+y$ from the neutral axis as shown. After bending through an angle $d\theta$, the length of the thin section changes from ℓ to $\ell + d\ell$.

There will be a plane in the cantilever, the so-called neutral axis, where no deformation occurs. Above the neutral axis $(+y)$, the horizontal "springs" are extended while below the neutral axis $(-y)$, the "springs" are compressed. For a rectangular cantilever with uniform cross-section, the neutral axis will coincide with the middle of the cantilever as drawn in Fig. 6.2. At any distance y from the neutral axis, the spring force f between atoms is thus translated into a moment that opposes the applied moment (Eq. (6.3)) and tends to restore the cantilever to its unbent shape.

Figure 6.3 shows an exaggerated view of the situation before and after an external force F is applied. After bending, any segment of the cantilever a distance y from the neutral axis will change its length from ℓ to $\ell + d\ell$. For any given y, Hook's Law requires that the internal forces df acting across the area element $dA = w_c dy$ must be proportional to the strain $d\ell/\ell$. Using the ideas discussed in Chapter 5, we thus have

$$\frac{df}{dA} = E\frac{d\ell}{\ell} \tag{6.4}$$

where E is the appropriate Young's modulus of the material from which the cantilever is fabricated. The deformation of the cantilever is assumed to be described by an arc that can be well described by a constant radius R. From Fig. 6.3, find that

$$d\theta = \frac{\ell}{R} = \frac{\ell + d\ell}{R + y}, \tag{6.5}$$

where R defines the radius of the bend to the neutral axis. From Eq. (6.5) it follows that

$$\frac{y}{R} = \frac{d\ell}{\ell}. \tag{6.6}$$

The internal force element df at y can then be written as

$$df = E\frac{y}{R}w_c dy. \tag{6.7}$$

In equilibrium, the applied moment generated by the force F applied to the end of the cantilever must equal the sum of all the internal moments along the beam. This means that at any point x, the following equality must hold

$$M_{\text{applied}} = F(L - x) = \int_{-t_c/2}^{+t_c/2} df \cdot y$$

$$= \int_{-t_c/2}^{+t_c/2} E\frac{y^2}{R}w_c \, dy = \frac{Ew_c t_c^3}{12}\frac{1}{R}. \tag{6.8}$$

Equation (6.8) relates the radius of the beam at point x to the applied force F. It is convenient to define the curvature κ as

$$\kappa \equiv \frac{1}{R} = \frac{12F(L - x)}{Ew_c t_c^3}. \tag{6.9}$$

Equation (6.9) can be compared to the formal mathematical definition of the curvature κ of an arbitrary line segment. The curvature is defined as the ratio of the change in the angle of a line tangent to the line segment for a given change in arc length. The relevant geometry to calculate the curvature of a line segment is sketched in Fig. 6.4.

Using the chain rule for derivatives, we have

$$\kappa \equiv \frac{d\theta}{d\ell} = \frac{d\theta}{dx}\frac{dx}{d\ell}. \tag{6.10}$$

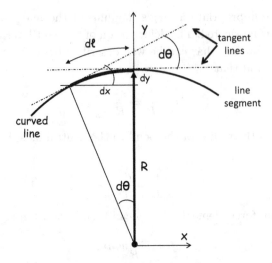

Fig. 6.4   The geometry required to define the radius of curvature κ of a line segment. Formally, $\kappa \equiv d\theta/d\ell$.

Example 6.1:   Calculate the curvature κ of a circle.   Use the formal definition given in Eq. (6.10) to define the curvature of a line segment. For a circle of radius R, the value of $d\ell$ integrated around the circle is the circumference $C = 2\pi R$. Over this distance $d\theta$ changes by an amount 2π. The curvature of a circle is therefore

$$\kappa = \frac{2\pi}{2\pi R} = \frac{1}{R}.$$

The two derivatives that appear in Eq. (6.10) must be evaluated. For sufficiently small $d\theta$, $d\ell$ can be written as

$$(d\ell)^2 = (dx)^2 + (dy)^2. \tag{6.11}$$

After some algebra, this leads to

$$\frac{dx}{d\ell} = \frac{1}{\sqrt{1 + \left(\frac{dy}{dx}\right)^2}}. \tag{6.12}$$

The factor $d\theta/dx$ in Eq. (6.10) can be evaluated using the fact that $\tan(d\theta) = \tan^{-1}(dy/dx)$. This gives

$$\frac{d\theta}{dx} = \frac{d}{dx}\left[\tan^{-1}\left(\frac{dy}{dx}\right)\right] = \frac{\frac{d}{dx}\left(\frac{dy}{dx}\right)}{1+\left(\frac{dy}{dx}\right)^2} = \frac{\frac{d^2y}{dx^2}}{1+\left(\frac{dy}{dx}\right)^2}. \tag{6.13}$$

Using Eqs. (6.12) and (6.13) in Eq. (6.10) finally gives the well-known result

$$\kappa \equiv \frac{d\theta}{d\ell} = \frac{\frac{d^2y}{dx^2}}{\left[1+\left(\frac{dy}{dx}\right)^2\right]^{\frac{3}{2}}} = \frac{1}{R}. \tag{6.14}$$

If the curvature of the cantilever is small, then dy/dx is small compared to 1. By combining Eq. (6.14) with Eq. (6.9), we have

$$\frac{1}{R} \simeq \frac{d^2y}{dx^2} = \frac{12F(L-x)}{Ew_ct_c^3}. \tag{6.15}$$

This differential equation can now be solved for the displacement of the cantilever y which is defined by $w(x)$ in Fig. 6.1.

$$w(x) \equiv \iint d^2y$$

$$= \iint \frac{12F(L-x)}{Ew_ct_c^3}dx^2 = \frac{12F}{Ew_ct_c^3}\left(\frac{Lx^2}{2} - \frac{x^3}{6}\right). \tag{6.16}$$

Equation (6.16) specifies the displacement of a cantilever from its equilibrium position when a force F is applied at the free end. When $x = L$, Eq. (6.16) simplifies to

$$w(L) \equiv q = \frac{4L^3}{Ew_ct_c^3}F. \tag{6.17}$$

Equation (6.17) clearly shows that the displacement of the cantilever q is proportional to the applied force F. Defining the spring constant k (in N/m) of the cantilever by $F = kq$, we finally have

$$k = \frac{Ew_ct_c^3}{4L^3}. \tag{6.18}$$

The slope of the cantilever will depend on x and is defined by

$$\Theta(x) = \frac{dy}{dx} = \frac{12F}{Ew_ct_c^3}\left(Lx - \frac{x^2}{2}\right). \tag{6.19}$$

When $x = L$, we have

$$\Theta(L) = \frac{6L^2}{Ew_ct_c^3}F. \tag{6.20}$$

From Eq. (6.20), the angle of deflection of the end of the cantilever is also proportional to the applied force F with a proportionality constant given by $2kL/3$. A convenient way to write $\theta(L)$ is

$$\Theta(L) = \frac{3}{2}\frac{q}{L}. \tag{6.21}$$

Lastly, many texts define the area moment I of the cantilever (units of m^4) as

$$I \equiv \int_{-t_c/2}^{+t_c/2} w_c y^2 dy = \frac{1}{12}w_ct_c^3. \tag{6.22}$$

With this definition, the standard equations derived above for a thin cantilever can be rewritten as

$$q \equiv y(L) = \frac{1}{3}\frac{L^3}{EI}F; \quad k = \frac{3EI}{L^3}; \quad \Theta(L) = \frac{L^2}{2EI}F. \tag{6.23}$$

Both q and Θ are proportional to F, the force applied to the end of the cantilever.

Example 6.2: Suppose you use a cantilever in an AFM experiment and the measured deflection is too small for the given force applied. You decide to increase the cantilever's sensitivity by a factor of 10. When searching for a new cantilever, what dimensions should it have and by how much would each dimension need to change to reach the new sensitivity?

The angular deflection is given by Eq. (6.20) as

$$\Theta(L) = \frac{6L^2}{Ew_ct_c^3}F.$$

Assuming a new cantilever will be made from the same material as the old, Young's modulus will remain the same. This leaves three

(Continued)

Example 6.2: (*Continued*)

possibilities to make Θ increase by a factor of 10 for the same applied force F: increase the cantilever length L, decrease the cantilever width w_c or decrease the cantilever thickness t_c.

The possibilities are listed in the table below.

Quantity changed	New value to increase sensitivity by a factor of 10	Comment
L	$L' = \sqrt{10}L \simeq 3.2L$	Length of cantilever increases by factor of ~ 3
w_c	$w_c' = 0.1 w_c$	Width of cantilever decreases by factor of ~ 10.
t_c	$t_c' = (0.1)^{1/3} t_c \simeq 0.46\, t_c$	Thickness of cantilever decreases by factor of $\sim 1/2$.

The required adjustment to L and w_c will produce a cantilever with markedly different dimensions from the original. A decrease in thickness might offer the best strategy to obtain the desired factor of ten increases in sensitivity.

The best strategy might be to find a cantilever with small changes in all three dimensions such that the overall gain in sensitivity is the required factor of 10.

6.2 The Cantilever as a Dynamic Beam

It is useful to derive a differential equation that describes the shape of a thin rectangular cantilever when it is subjected to a static load applied at the end of the cantilever as shown in Fig. 6.5(a). We seek a systematic treatment of this problem that can be easily extended to the situation when the cantilever is driven sinusoidally by an external force, a situation important for dynamic AFM. When a static distributed load $P(x)$ is applied along the beam, the beam will deflect until every part of the beam has no net force acting on it. If we treat the cantilever as a classical thin beam, the required equation of motion can be derived by applying the results learned in the study of the mechanics of materials. The results obtained from this approach can be extended to analyze cantilever vibrations.

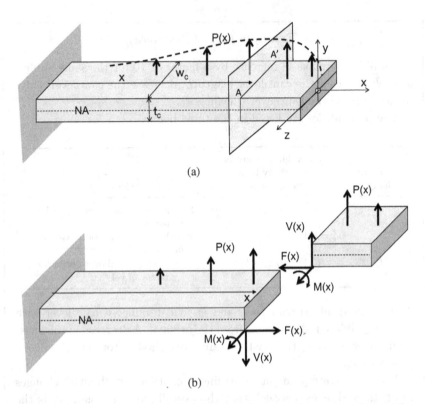

(a)

(b)

Fig. 6.5 In (a), an arbitrary load $P(x)$ distributed along a thin rectangular cantilever pinned at one end. The neutral axis (NA) is indicated by the dotted line and lies along the midpoint of the cantilever. In (b), the cantilever is split into two sections, one section to the left of the plane AA$'$ and one section to the right. Three internal resultants (two forces and one moment) are important: (i) an internal shear force $V(x)$, (ii) an internal axial force $F(x)$, and (iii) an internal bending moment $M(x)$. Each internal resultant developed by the left cantilever section must be balanced by an equivalent internal resultant developed by the right cantilever section.

Consider an arbitrary plane AA$'$ that intersects the cantilever as shown in Fig. 6.5(a). Equilibrium basically means that the part of the beam to the left of this plane must exert equal but opposite forces on the remainder of the beam to the right of the plane to prevent it from accelerating.

As shown in Fig. 6.5(b), three internal resultants (two forces and one moment) are important: (i) an internal shear force $V(x)$, (ii) an internal axial force $F(x)$ and (iii) an internal bending moment $M(x)$. The magnitude of these three quantities will depend on the location of the plane AA$'$ and therefore must depend on x. Since we are concerned about the transverse deflection of the beam, the axial (pulling) force F along the beam is not relevant.

There are important relations that can be derived that interconnect these three quantities. Assume there is a distributed load $P(x)$ (units of N/m) along the beam. A resulant force can be defined which has the same effect on a rigid body as the original distributed load. The location $\bar{x}$ of the resultant force is given by the centroid of $P(x)$ defined by

$$\bar{x} = \frac{\int_0^L x\, P(x)\, dx}{\int_0^L P(x)\, dx}. \tag{6.24}$$

Consider a segment of the beam of infinitesimal length Δx at a position x as shown in Fig. 6.6. Suppose both $V(x)$ and $M(x)$ act on the left face of this beam segment. On the right side, the relevant resultants are $V(x+\Delta x)$ and $M(x + \Delta x)$. A force balance along this segment implies that

$$V(x + \Delta x) = V(x) + P(x)\Delta x. \tag{6.25}$$

As $\Delta x \to 0$, it follows that

$$\frac{dV}{dx} = P(x). \tag{6.26}$$

Thus if the loading force per unit length is known as a function of position, the variation in the internal shear force is also known.

In a similar way, if a balance of moments is calculated about the point on the right-most face marked by the black dot in Fig. 6.6, you must

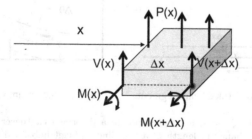

Fig. 6.6 A cantilever segment of length Δx showing the internal shear force $V(x)$ and an internal bending moment $M(x)$. The diagram is useful to derive how V and M vary with position.

have

$$M(x + \Delta x) = M(x) + V(x)\Delta x + P(x)\Delta x \left(\frac{\Delta x}{2}\right). \qquad (6.27)$$

As $\Delta x \to 0$, the term quadratic in Δx becomes vanishingly small and we have

$$\frac{dM}{dx} = V(x). \qquad (6.28)$$

The shear force $V(x)$ and bending moment $M(x)$ must be formally connected to stresses and strains which allow a determination of the deformation of the cantilever. The classical Bernoulli-Euler assumption, valid for the bending of a thin, homogeneous beam, is that a section of the beam located along the neutral axis is under no strain. This means that the length of a small segment of the beam located on the neutral axis remains unchanged after bending.

This situation is sketched in Fig. 6.7 which illustrates that a line segment (imbedded in the beam) has a uniform length, independent of y, before bending. After bending to produce a **local** radius of curvature R, the line segment is stretched or compressed, depending on its location with respect to the neutral axis. The change in length at a distance y from the neutral axis divided by the original segment length is the strain ε (no units) and is given by

$$\varepsilon(x, y) = \lim_{\Delta x \to 0} \frac{(R - y)\Delta\theta - R\Delta\theta}{R\Delta\theta} = \frac{-y}{R}. \qquad (6.29)$$

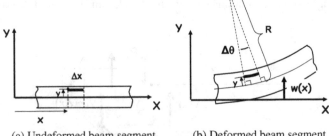

(a) Undeformed beam segment   (b) Deformed beam segment

Fig. 6.7   Calculating the stress and strain in a deformed cantilever beam requires a calculation of the change in length of a thin line segment imbedded in the cantilever. Before deformation, the thin segment has a uniform length Δx independent of y. After deformation, the length of the thin segment depends on y. When $y = 0$ (the neutral axis), the line segment is assumed to have its original length, even after deformation.

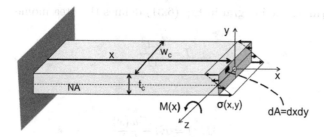

Fig. 6.8   A schematic of how the stress $\sigma(x, y)$ varies in magnitude and direction along a cross-sectional face of a thin rectangular cantilever fixed at one end.

According to Hook's Law, the resulting strain σ (units of Pa) is proportional to the stress multiplied by Young's modulus of the beam

$$\sigma(x, y) = E\left(\frac{-y}{R}\right) = -Ey\frac{d^2w}{dx^2}, \tag{6.30}$$

where Eq. (6.15) has been used to relate the radius of curvature R to the second derivative of the deflected beam. This result is valid for small deflections.

With an expression for the stress, the internal moment $M(x)$ can be calculated at x by integrating the stress over the thickness of the cantilever. The relevant geometry is shown in Fig. 6.8. Above the neutral axis, the cantilever suffers a compressive stress as shown by the direction of σ. Below the neutral axis, the stress is tensile and tends to stretch the cantilever outward. Along the neutral axis, the stress is zero. If the cantilever has a non-zero moment at x, it must be due to the collective action of the stress over the cantilever's cross-sectional area. This can be calculated from the force at each point $(\sigma(x, y)dy\,dz)$ multiplied by the moment arm y integrated over the cantilever:

$$\begin{aligned}
M(x) &= \int_{z=-w_c/2}^{w_c/2} \int_{y=-t_c/2}^{t_c/2} \sigma(x, y)y\,dy\,dz \\
&= \int_{z=-w_c/2}^{w_c/2} \int_{y=-t_c/2}^{t_c/2} Ey^2\frac{d^2w(x)}{dx^2}dy\,dz \\
&= E\frac{d^2w(x)}{dx^2}\int_{z=-w_c/2}^{w_c/2} \int_{y=-t_c/2}^{t_c/2} y^2dy\,dz. \tag{6.31}
\end{aligned}$$

Performing the integral in Eq. (6.31) defines the area moment

$$I = \frac{w_c t_c^3}{12}.$$ (6.32)

This gives

$$M(x) = EI\frac{d^2 w(x)}{dx^2}.$$ (6.33)

Since Eq. (6.28) relates the derivative of $M(x)$ to $V(x)$, we also have

$$V(x) = EI\frac{d^3 w(x)}{dx^3}.$$ (6.34)

Finally, according to Eq. (6.26), the derivative of $V(x)$ is related to the distributed load $P(x)$, which gives

$$EI\frac{d^4 w(x)}{dx^4} = P(x).$$ (6.35)

Equation (6.35) is the desired result and must then be solved for $w(x)$ subject to boundary conditions at $x = 0$ and $x = L$.

Example 6.3: Solve the differential equation (Eq. (6.35)) for the deformation of a cantilever beam when a point force F is applied to the free end.

Since there is no distributed load, $P(x) = 0$ and Eq. (6.35) becomes

$$EI\frac{d^4 w(x)}{dx^4} = 0.$$

(*Continued*)

Example 6.3: (*Continued*)

Integrating this equation four times gives us the following sequence of equations

$$\frac{d^3w(x)}{dx^3} = c_1 \qquad \frac{d^2w(x)}{dx^2} = c_1x + c_2,$$

$$\frac{dw(x)}{dx} \equiv \theta(x) = \frac{1}{2}c_1x^2 + c_2x + c_3,$$

$$w(x) = \frac{1}{6}c_1x^3 + \frac{1}{2}c_2x^2 + c_3x + c_4.$$

The four constants of integration must be determined by the boundary conditions at $x = 0$ and $x = L$. At $x = 0$, the cantilever is pinned and the displacement and slope must equal zero. This gives

$$w(x = 0) = 0, \qquad \text{(I)}$$
$$\theta(x = 0) = 0. \qquad \text{(II)}$$

At $x = L$, there is no internal moment M and the shear force is just opposite the applied force F. From Eqs. (6.33) and (6.34), this gives

$$M(x) = EI\frac{d^2w(L)}{dx^2} = 0, \qquad \text{(III)}$$

$$V(x) = EI\frac{d^3w(L)}{dx^3} = -F. \qquad \text{(IV)}$$

Equation (I) implies $c_4 = 0$; Equation (II) implies $c_3 = 0$; Equation (IV) implies that

$$c_1 = -\frac{F}{EI}.$$

Lastly, at $x = L$, Eq. (III) gives the condition that

$$c_1L + c_2 = 0.$$

From which it follows that

$$c_2 = \frac{FL}{EI}.$$

(*Continued*)

Example 6.3:   (*Continued*)

With the four constants of integration determined, the deflection of the free end of the beam, $w(x = L)$, can be evaluated as

$$w(x = L) \equiv q = \frac{1}{3}\frac{L^3}{EI}F,$$

which is identical to the result obtained in Eq. (6.23) above. Writing

$$F = kq$$

gives an expression for the spring constant of the cantilever k as

$$k = \frac{3\,EI}{L^3},$$

which also agrees with the results given in in Eq. (6.23) above.

6.3   The Cantilever as a Vibrating Beam

The extended discussion presented above allows the differential equation for a vibrating cantilever to be written down in a straightforward way. If the cantilever is vibrating, then the deflection along the cantilever $w(x)$ must be written as $w(x, t)$ because the deflection is no longer independent of time. The force acting on any beam segment must then equal mass times acceleration according to Newton's First Law of motion. Equation (6.35) must be modified to include the acceleration of the cantilever with the result that the differential equation now governing the motion of the beam can be written as

$$EI\frac{\partial^4 w(x,t)}{\partial x^4} = -\rho_L \frac{\partial^2 w(x,t)}{\partial t^2}, \tag{6.36}$$

where ρ_L is the mass density per unit length, calculated by multiplying ρ_o, the mass density of the cantilever per unit volume, by the cantilever's cross-sectional area. The bending deflection can be decomposed into functions $\Phi_j(x)$ which describes the various mode shapes of the oscillation times a sinusoidal function of time which describes the frequency of each vibrating mode. Thus we posit a solution of Eq. (6.36) containing an infinite number of modes j of the form

$$w(x, t) = \sum_{j=1}^{\infty} c_j \Phi_j(x)\, e^{\pm i\omega t}, \tag{6.37}$$

where as usual, $i = \sqrt{-1}$. When Eq. (6.37) is substituted into Eq. (6.36), the following equation results

$$\frac{\partial^4 \Phi_j(x)}{\partial x^4} - \left(\frac{\alpha_j}{L}\right) \Phi_j(x) = 0, \tag{6.38}$$

where

$$\left(\frac{\alpha_j}{L}\right)^4 = \frac{\rho_L \omega_j^2}{EI}. \tag{6.39}$$

Equation (6.38) is solved by applying appropriate boundary conditions for a free cantilevered beam pinned at one end. These boundary conditions are:

(i) at $x = 0$, the displacement of the cantilever is zero ($\Phi_j(x = 0) = 0$);
(ii) at $x = 0$, the angular deflection of the cantilever is zero ($d\Phi_j(x = 0)/dx = 0$);
(iii) at $x = L$, the torque on the free end of the cantilever is zero ($d^2\Phi_j(x = L)/dx^2 = 0$); and
(iv) at $x = L$, the force acting on the free end of the cantilever is zero ($d^3\Phi_j(x = L)/dx^3 = 0$).

Applying the boundary conditions yields a characteristic transcendental equation

$$\cos(\alpha_j)\cosh(\alpha_j) + 1 = 0 \tag{6.40}$$

that must be solved numerically to identify the allowed values of α_j. Each value of α_j satisfying Eq. (6.40) then specifies the eigenfrequencies given by Eq. (6.39) for each eigenmode j. The solutions for Eq. (6.40) are well known and Table 6.1 lists the first three values. Each solution corresponds to a discrete eigenfrequency ω_j as specified by Eq. (6.39) and a particular modal shape as shown in Table 6.1. [raman09].

The above discussion is focused on thin cantilevers of rectangular cross-section. As cantilevers become smaller, the effect of tip location and added tip mass becomes an important factor that requires a modification to the classical discussion outlined above.

It is useful to express the eigenfrequencies $f_j = \omega_j/2\pi$ (in Hz) from Eq. (6.39) in a form that depends solely on the cantilever's dimensions and materials properties. For a cantilever with a rectangular cross-section

Table 6.1   The first three modes for a rectangular cantilever are specified by the separation constant α_j. The shape of the first three eigenmodes is also given. (Reproduced with permission from [spletzer07]).

j	1	2	3
α_j	1.8751	4.6941	7.8548

of length L, width w_c, thickness t_c, and mass $m_c = \rho_o L w_c t_c$, Eq. (6.39) becomes

$$f_j = \frac{1}{2\pi} \frac{\alpha_j^2}{2\sqrt{3}} \frac{t_c}{L^2} \sqrt{\frac{E}{\rho_o}} = \frac{1}{2\pi} \frac{\alpha_j^2}{2\sqrt{3}} \sqrt{\frac{E\, w_c}{m_c}} \left(\frac{t_c}{L}\right)^3. \qquad (6.41)$$

Equation (6.41) indicates the various resonant mode frequencies are in the ratio $f_1 : f_2 : f_3 : \ldots = f_0 : 6.3f_0 : 17.5f_0 \ldots$. Recent trends in AFM make use of these higher eigenmodes to enhance imaging and to determine local material properties. These developments are discussed in part II of these lecture notes.

Lastly it is worth briefly mentioning that in addition to cantilever oscillations at the eigenfrequencies discussed above, it is also possible for the cantilever to oscillate at frequencies equal to integer multiples of each eigenfrequency given by Eq. (6.41). [crittenden05], [shaik14]. These higher harmonics are especially useful for reconstructing the periodic force that the tip experiences in dynamic AFM applications and will be discussed in more detail in Part II of these lecture notes.

6.4   Thermal Vibrations of the Cantilever

The well-known equipartition theorem from thermodynamics states that for a system in thermal equilibrium with a reservoir at a temperature T, there is $1/2 k_B T$ of thermal energy associated with each independent degree of freedom. The implications of the equipartition theorem, which provides considerable insight into the ideal gas law describing non-interacting gas atoms, has broad implications when applied to a cantilever surrounded by a gas held at a temperature T.

The mathematics required to describe a randomly fluctuating cantilever is quite different from the geometry-based mathematics discussed in the previous sections of this chapter. A general discussion that provides the necessary background is presented in what follows.

6.4.1 Describing random fluctuations of cantilever displacement

When a cantilever is buffeted by impacts from surrounding gas molecules, a random excitation of the cantilever occurs that causes it to twitch with time. Since the gas atom impacts are random, the resulting displacement of the cantilever $q(t)$ as a function of time becomes a random variable. A random process is deemed stationary if all statistics are invariant with respect to a time shift, a condition satisfied by the fluctuations of the cantilever. It is interesting to devise a framework to characterize the degree of "randomness" of the fluctuations. Such a context can be found in probability theory devised to measure random signals.

The variation of random fluctuations is difficult to determine because the physical events responsible for the fluctuations are essentially unpredictable and defy analytical description. One way to proceed might be to measure the mean value of the cantilever's position over time. Since the fluctuations occur about a well-defined average position, the mean value of $q(t)$ over time will yield only a single number, $\langle q \rangle$. Not much information is obtained in this way since there are an infinite collection of numbers that give the same average value. Another approach might be to measure the root-mean-square (rms) value of q by measuring N values of q at different times and calculating the average value of the squared difference from the mean

$$q_{\mathrm{rms}} = \sum_{i=1}^{N} \sqrt{\frac{(q(t_i) - \langle q \rangle)^2}{N}}. \tag{6.42}$$

By calculating the square of the deviation from the mean, a more representative measure of the fluctuations will result than just the mean value itself. The value of q_{rms} accurately captures the extent of the fluctuations from the cantilever's mean position.

However, this approach is also limited, because different time histories of the cantilever's position might have the same rms value, but still differ in important ways. As an example, consider two cantilevers. Cantilever 1 might contain long time-scale variations compared to Cantilever 2 yet still have the same rms value. Or Cantilever 1 could exhibit significantly more fluctuations in the high-frequency (short time-scale) range while Cantilever 2 could have fluctuations of equal magnitude at all frequencies, yet both cantilevers could still have the same numerical q_{rms} value. How can such information about the differences in fluctuations between these two cantilevers be captured? These important differences only become evident if the fluctuating cantilever is further analyzed in the frequency domain.

In what follows, we assume the cantilever fluctuation $q(t)$ is transduced by a photodiode, so that the fluctuations are accurately represented by a voltage signal. As discussed in Chapter 7, Sec. 7.3, the cantilever deflection q (in m) is related to the voltage from the photodiode $V(t)$ by the photodiode sensitivity S_z (in units of m/V) by

$$q(t) = S_z V(t). \tag{6.43}$$

This conversion of deflection to voltage allows us to discuss correlations in voltage signals that fluctuate in time, a topic that is usually covered in undergraduate electrical and mechanical engineering courses [lathi68].

Important statistical information is contained in the autocorrelation function $R_{t,t+\Delta t}$ which tells how similar a signal is to itself after a specified time shift $\Delta t = \tau$. The auto-correlation function enjoys multiple definitions, depending on the circumstances. For a signal of infinite duration, $R_{t,t+\tau}$ is often defined as

$$R_{t,t+\tau}(\tau) = \lim_{T \to \infty} \frac{1}{T} \int_{-T/2}^{+T/2} q^*(t)q(t+\tau)\, dt \quad \text{(infinite duration)}. \tag{6.44}$$

The quantity q^* in Eq. (6.44) is the complex conjugate of q; for real q, then $q^* = q$. The factor $1/T$ appearing in front of the integral provides the rationale for a "power" designation of $R_{t,t+\tau}$ as is often found in the literature.

Other definitions commonly encountered are for a signal of finite duration ($t_1 < t < t_2$)

$$R_{t,t+\tau}(\tau) = \int_{t_1}^{t_2} q^*(t)q(t+\tau)dt \quad \text{(finite duration)} \tag{6.45}$$

which gives rise to an "energy" designation of the autocorrelation function.

If the signal is periodic with period T, one often finds the definition

$$R_{t,t+\tau}(\tau) = \frac{1}{T} \int_{t_1}^{t_1+T} q^*(t)q(t+\tau)dt \quad \text{(periodic)} \qquad (6.46)$$

where t_1 represents any arbitrary time. This definition again recovers the "power" designation of the autocorrelation function.

The utility of $R_{t,t+\tau}$ should be readily apparent. For a fluctuating signal $q(t)$, there might be a peak in $R_{t,t+\tau}$ for $\tau < t_o$, where t_o represents a small time shift. Such a localized peak in $R_{t,t+\tau}$ near $\tau = 0$ would indicate a correlated signal over short times. If $R_{t,t+\tau}$ tends to zero as τ increases, this would indicate that $q(t)$ is randomly fluctuating and there is no correlation between the signal at time t and time $t + \tau$. If $q(t)$ contains a contribution that varies sinusoidally at some frequency f, then $R_{t,t+\tau}$ should have a peak whenever $\tau = 1/f$.

Problem 6.5: Calculate the autocorrelation function of $q(t) = A \sin(\omega t)$. We must first define $q(t)$ and $q(t+\tau)$. In this case, these assignments are easy to make:

$$q(t) = A \sin[\omega t] \quad q(t+\tau) = A \sin[\omega(t+\tau)].$$

Since the signal is periodic, we use the form of the autocorrelation function given in Eq. (6.46):

$$R_{t,t+\tau}(\tau) = \frac{1}{T} \int_{t_1}^{t_1+T} q^*(t)q(t+\tau)dt$$

$$= \frac{A^2}{T} \int_0^{T=2\pi/\omega} \sin(\omega t) \sin(\omega(t+\tau))dt.$$

Let $\alpha = \omega t$ and $\beta = \omega\tau$.

(Continued)

Problem 6.5: (*Continued*)

We have $\sin(\alpha)\sin(\alpha+\beta) = \frac{1}{2}[\cos(\beta) - \cos(2\alpha+\beta)]$

$$R_{t,t+\tau}(\tau) = \frac{A^2}{2T}\int_0^{T=2\pi/\omega}\cos(\omega\tau)dt - \frac{A^2}{2T}\int_0^{T=2\pi/\omega}\cos(2\omega t+\omega\tau)dt$$

$$= \frac{A^2}{2T}\cos(\omega\tau)[T-0] - \frac{A^2}{2T}\left[\frac{\sin(2\omega t+\omega\tau)}{2\omega}\right]_0^{\frac{2\pi}{\omega}}$$

$$= \frac{A^2}{2}\cos(\omega\tau) - \frac{A^2}{2T}\left[\frac{\sin\left(2\omega\left(\frac{2\pi}{\omega}\right)+\omega\tau\right)}{2\omega} - \frac{\sin(\omega\tau)}{2\omega}\right]$$

$$= \frac{A^2}{2}\cos(\omega\tau) - \frac{A^2}{2T}\left[\frac{\sin(\omega\tau)}{2\omega} - \frac{\sin(\omega\tau)}{2\omega}\right]$$

$$= \frac{A^2}{2}\cos(\omega\tau) = \frac{A^2}{2}\cos\left(2\pi\frac{\tau}{T}\right)$$

The autocorrelation function of a sine wave with amplitude A and period T is therefore periodic in the time shift τ with a period of $\tau = T$ and an amplitude of $A^2/2$.

6.4.2 *The spectral power density*

For a random signal of infinite duration, it is useful to spectrally analyze the autocorrelation function $R_{t,t+\tau}$ given by Eq. (6.44) since then any periodic fluctuations in the system can be readily assessed. The spectral content of $R_{t,t+\tau}$ can be obtained by performing a Fourier decomposition to calculate the power spectral density (PSD) of the stationary fluctuation $q(t)$ which is defined by

$$PSD(\omega) \equiv \int_{-\infty}^{+\infty} R_{t,t+\tau}(\tau)e^{i\omega\tau}\,d\tau. \qquad (6.47)$$

Within the context of an AFM application to characterize thermal noise, the power spectral density defined in Eq. (6.47) provides a proportional relationship between a measure of the square of the cantilever's thermal displacement $q(t)$ and a measure of the amplitude of the PSD at different frequencies. Unfortunately, there are many conventions regarding the normalization between the squared amplitude

$R_{t,t+\tau}$ and the PSD. The many numerical PSD estimators available in software packages must be evaluated and tested with this caveat in mind [press92]. Without attention to this detail, the calculated PSD will be difficult to understand.

According to Eq. (6.47), the PSD is defined as the Fourier transform of the autocorrelation function. The power spectral density of a fluctuating signal $q(t)$ will therefore have units of $[(q)^2 \times \sec/\text{rad}] = [(q)^2/\text{Hz}]$. Thus if the fluctuation in the cantilever position $q(t)$ is measured in units of picometers (pm), then the corresponding units of PSD(ω) will be $[(\text{pm})^2/\text{Hz}]$.

Alternatively, if the cantilever fluctuation is measured by monitoring the output *voltage* of the photodiode, then q will be measured in volts and the units of PSD(ω) will be $[(\text{V})^2/\text{Hz}]$. When voltages are the measured quantity, graphs are often presented in which the square root of the PSD is plotted as a function of frequency. You can recognize this convention by the units which will be $[\text{V}/\sqrt{\text{Hz}}]$.

While the formal definition of PSD(ω) is clear, the best way to numerically calculate PSD(ω) from time series data requires attention to technical details that will distract us from the present discussion [press92]. Suffice it to say that a plot of the magnitude of PSD vs. ω provides a visual characterization of those frequencies which are dominant in a measured signal since the PSD(ω) measures the distribution of power over a broad frequency range.

Figure 6.9 shows representative plots of PSD(ω) for different situations that include a 10 mV peak-to-peak random noise signal, a well-defined sinusoidal wave with constant amplitude added to a 10 mV peak-to-peak random noise signal, and a clipped sine wave added to a 10 mV peak-to-peak random noise signal.

By using an analytical model for a cantilever (e.g., the damped harmonic oscillator), a theoretical prediction of PSD(ω) can be made and compared to experimental measurements to extract parameters of interest. A further advantage to determining the PSD(ω) is that periodic noise in the voltage from say the photodiode (like 60 Hz hum due to a bad electrical ground) is easy to identify.

There are two important results related to the calculation of PSD(ω) that are relevant to our discussion. One is derived from the cross-correlation theorem and relies on the Fourier transform of $q(t)$ in terms of its Fourier

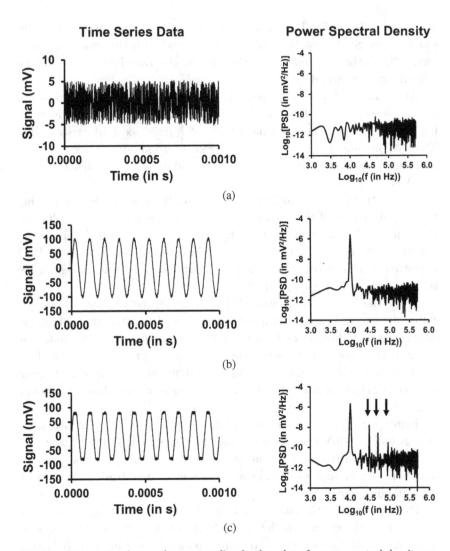

Fig. 6.9　Time series data and corresponding log-log plot of power spectral density vs. frequency (in Hz) for (a) 10 mV peak-to-peak random noise signal, (b) sine wave + random noise (SNR = 10:1), and (c) clipped sine wave plus random noise. The arrows in (c) mark the emergence of higher frequency components in the distorted sine wave. The random noise in all cases contributes a nearly constant signal in the PSD that spans all frequencies.

transform pair $Q(\omega)$

$$
\begin{aligned}
R_{t,t+\tau}(\tau) &= \int_{-\infty}^{+\infty} q^*(t)q(t+\tau)dt \\
&= \int_{-\infty}^{+\infty} \left[\left(\int_{-\infty}^{+\infty} Q^*(\omega)e^{i\omega t}d\omega \right) \left(\int_{-\infty}^{+\infty} Q(\omega')e^{-i\omega'(t+\tau)}d\omega' \right) \right] dt \\
&= \int_{-\infty}^{+\infty} \int_{-\infty}^{+\infty} \int_{-\infty}^{+\infty} Q^*(\omega)Q(\omega')e^{i(\omega-\omega')t}e^{-i\omega'\tau}dt\,d\omega\,d\omega' \\
&= \int_{-\infty}^{+\infty} \int_{-\infty}^{+\infty} Q^*(\omega)Q(\omega')e^{-i\omega'\tau} \left[\int_{-\infty}^{+\infty} e^{i(\omega-\omega')t}dt \right] d\omega\,d\omega' \\
&= \int_{-\infty}^{+\infty} \int_{-\infty}^{+\infty} Q^*(\omega)Q(\omega')e^{-i\omega'\tau}\delta(\omega-\omega')d\omega\,d\omega' \\
&= \int_{-\infty}^{+\infty} Q^*(\omega)Q(\omega)e^{-i\omega\tau}d\omega \\
&= \int_{-\infty}^{+\infty} |Q(\omega)|^2 e^{-i\omega\tau}d\omega.
\end{aligned}
\tag{6.48}
$$

Equation (6.48) allows a convenient calculation of PSD(ω) since we can now write $R_{t,t+\tau}$ in terms of $Q(\omega)$

$$
\begin{aligned}
PSD(\omega) &\equiv \int_{-\infty}^{+\infty} R_{t,t+\tau(\tau)}e^{i\omega\tau}d\tau \\
&= \int_{-\infty}^{+\infty} \left[\int_{-\infty}^{+\infty} |Q(\omega')|^2 e^{-i\omega'\tau}d\omega' \right] e^{i\omega\tau}d\tau \\
&= \int_{-\infty}^{+\infty} |Q(\omega')|^2 \left[\int_{-\infty}^{+\infty} e^{-i(\omega-\omega')\tau}d\tau \right] d\omega' \\
&= \int_{-\infty}^{+\infty} |Q(\omega')|^2 \left[\delta(\omega-\omega') \right] d\omega' \\
&= |Q(\omega)|^2
\end{aligned}
\tag{6.49}
$$

The second important result, known as Parseval's Theorem (also called Pancheral's Theorem), establishes an equality between integrals over $q(t)$ and $Q(\omega)$.

$$\int_{-\infty}^{+\infty} |q(t)|^2 dt = \int_{-\infty}^{+\infty} |Q(\omega)|^2 d\omega. \tag{6.50}$$

Equation (6.50) simply states that the average power in a signal $[q^2(t)]$ is equal to the sum of the average power in each of its Fourier components $[Q(\omega)]^2$.

6.4.3 *A few signal processing applications*

To finish this discussion, it is useful to build intuition and discuss a few standard examples that illustrate why knowledge of PSD(ω) is useful within the context of voltage signals.

The first example is a quantification of the noise in an electronic circuit. Once the PSD(ω) is measured, it can be used to predict the noise level in a system. Suppose the PSD of an electronic circuit is known as a function of ω. If a voltage signal is measured with an instrument that accurately measures the signal over a limited (but known) frequency range or bandwidth, then the rms noise voltage will be given by integrating over the bandwidth

$$V_{\text{rms}}^2 = \int_{\omega_{\text{low}}}^{\omega_{\text{high}}} \text{PSD}(\omega)\, d\omega. \tag{6.51}$$

To be specific, consider the case of 'white' noise. White noise has a constant power-spectral density PSD(ω) which is independent of frequency. If the white noise signal from a circuit is measured to have a constant value P_0, i.e., PSD(ω) $= P_0 = 1 \times 10^{-10}$ V^2/Hz. Then using a measuring instrument like an oscilloscope with a bandwidth of say $B = 10\,\text{MHz}(\omega_{\text{low}} \simeq 0; \omega_{\text{high}} \simeq 2\pi f_{\text{high}}; B \simeq f_{\text{high}} - f_{\text{low}})$, you might expect to measure an rms noise voltage given by

$$V_{\text{rms}}^2 = P_0 B. \tag{6.52}$$

Thus $V_{\text{rms}} = \sqrt{P_0 B} = 32\,\text{mV}$. A second common application of the power spectral density is to estimate the effect of an electronic filter on reducing the noise level in a signal. Electronic noise filtering is a common way to decrease noise and improve the signal to noise ratio (SNR). Suppose a filter has a particular linear-response function specified by

$$H(\omega) = \frac{V_{\text{out}}(\omega)}{V_{\text{in}}(\omega)} \tag{6.53}$$

(note that we only specify the absolute value of H here). The action of the filter at a particular frequency ω in a system with a known PSD(ω) can then be specified as

$$V_{\text{out}}^2(\omega) = H^2(\omega)\text{PSD}(\omega). \tag{6.54}$$

For a first-order low-pass filter, a common response function is often written as

$$H(\omega) = \frac{1}{\sqrt{1 + \left(\frac{\omega}{\omega_{3dB}}\right)^2}}, \tag{6.55}$$

where ω_{3dB} is some cutoff frequency that specifies the attenuation of the filter at high frequencies. It follows that the rms noise after filtering will be given by

$$V_{\text{out}}^2 = \int_o^\infty H^2(\omega)\text{PSD}(\omega)\,d\omega. \tag{6.56}$$

If the source of noise is white, then $\text{PSD}(\omega) = P_0$ (in V^2/Hz) where P_0 is a constant independent of frequency. We then have

$$V_{\text{out}}^2 = P_0 \int_o^\infty H^2(\omega)\,d\omega = P_0 \int_o^\infty \left[1 + \left(\frac{\omega}{\omega_{3dB}}\right)^2\right]^{-1} d\omega. \tag{6.57}$$

The integral in Eq. (6.57) can be evaluated to give

$$I = \int_o^\infty \frac{dx}{1 + a^2 x^2} = \frac{1}{a}\tan^{-1}(ax)\big|_0^\infty = \frac{1}{a}\left[\frac{\pi}{2}\right] \tag{6.58}$$

and we finally have

$$V_{\text{out}}^2 = \omega_{3dB}\left[\frac{\pi}{2}\right]P_0. \tag{6.59}$$

By comparison to Eq. (6.52), an effective bandwidth of the filter can be defined. The quantity $\pi\omega_{3dB}/2$ in Eq. (6.59) is often called the effective noise bandwidth (ENBW) of the filter specified by $H(\omega)$ defined in Eq. (6.55). The ENBW is the equivalent bandwidth of a rectangular filter that passes the same power of white noise as the window $H(\omega)$. This definition of ENBW holds only if white noise is dominant.

6.5   Chapter Summary

A summary of the mechanical properties of cantilevers has been presented with an aim to identify the important factors controlling the bending and vibration of a thin rectangular cantilever beam to its geometric dimensions. The goal is to understand the principles underlying standard formulas used by experimentalists to estimate spring constants and eigenfrequencies of AFM cantilevers. While experts have published many papers to further refine this understanding, it is worthwhile to appreciate the underlying physics and mechanics of the simple classic models that have been developed.

A discussion of a cantilever vibration due to random thermal impact by gas molecules has also been included, since the mathematics required to describe this random process is considerably different than the differential equations required to describe cantilever bending and vibration. By measuring the thermal fluctuation of the cantilever, it is possible to calibrate the spring constant in a simple way. This calibration will be discussed in more detail in Chapter 8.

A discussion of a cantilever driven at some frequency is not included, since this topic in all its detail will be covered in Part II of these lecture notes when discussing dynamic AFM.

6.6   Further Reading

References to the Original Literature:

[crittenden05] S. Crittenden, A. Raman and R. Reifenberger,. "Probing attractive forces at the nanoscale using higher harmonic dynamic force microscopy", Phys. Rev. **B72**, 235422–35 (2005).

[lathi68] B.P. Lathi, *Random Signals and Communication Theory* (International Textbook Co., Scranton, PA), 1968.

[press92] W.H. Press, S.A. Teukolsky, W.T. Vettering, and B.P. Flannery, Numerical Recipes in C, The Art of Scientific Computing, 2nd Ed. (Cambridge University Press, Cambridge UK, 1992), see Chapter 13.

[raman09] "Cantilever dynamics and nonlinear effects in Atomic Force Microscopy," A. Raman, R. Reifenberger, J. Melcher and R. Tung, a book chapter in *Noncontact Atomic Force Microscopy Vol. 2*, Eds. S. Morita, F.J. Giessibl, and R. Wiesendanger (Springer, Berlin, 2009), pgs. 361–91.

[spletzer07] M. Spletzer, A. Raman and R. Reifenberger, "Elastometric sensing using higher flexural eigenmodes of microcantilevers", Appl. Phys. Lett. **91**, 184103–06 (2007).

[shaik14] N.H. Shaik, R.G. Reifenberger, A.Raman, "Microcantilevers with embedded accelerometers for dynamic atomic force microscopy", Appl. Phys. Lett. **104**, 083109 (2014).

Further reading:

[wortman64] J.J. Wortman and R.A. Evans, "Young's Modulus, Shear Modulus and Posison's Ratio in Silicon and Germanium", J. App. Phys. **36**, 153 (1964).

[lübbe13] J. Lübbe et al., "Thermal noise limit for ultra-high vacuum noncontact atomic force microscopy", Beilstein J. Nanotechnol. **4**, 32–44 (2013).

6.7 Problems

1. A web site listing Si cantilevers provides a number of choices, but only provides the dimension (length L, width w_c and thickness t_c) for each cantilever. Complete the table below. The density of Si is $2330 \, \text{kg/m}^3$ and the elastic modulus of Si(100) is $130 \, \text{GPa}$.

Cantilever	L (μm)	w_c (μm)	t_c (μm)	k (N/m)	Resonance Frequency (Hz)
A	460	50	2		
B	230	40	3		
C	125	35	4		

2. What is the mass of cantilever C listed in the table given in Problem 1 above?

3. Suppose you want to fabricate a microcantilever from a Si wafer that deflects 1 nm when 1 nN of force is applied at the end. Most (if not all) Si cantilevers are made from ⟨100⟩ oriented Si wafers.

   (a) Suppose you already have a reliable fabrication process to produce cantilevers that are 30 μm wide. What should be the ratio of the thickness to the length of the microcantilever to achieve the desired performance of 1 nm deflection when 1 nN of force is applied?

   (b) If the minimum thickness of the cantilever you can reliably fabricate is 2 μm, what should be the length of cantilever to meet the desired specification?

   (c) Due to fabrication issues, if the uncertainty in the thickness of the cantilever is ±10%, what is the corresponding uncertainty in the spring constant?

4. What is the resonant frequency of the 3rd mode of cantilever B listed in the table given in Problem 1 above?
5. Using the results of Example 6.3 from Chapter 6, write down the equation for the deformation $w(x)$ of a cantilever beam of length L when a point force F is applied to the free end. Make a plot of $w(x)$ for cantilever A in Problem 1 when a force of 1 nN is applied to the end of the cantilever.

   (a) What is the resulting cantilever deflection at the free end?
   (b) Roughly what distance from the fixed end is required before the cantilever will deflect more than 1 nm? Express your answer as a fraction of L.

6. When a point force F is applied to the end of a cantilever, the cantilever bends and takes on shape that can roughly be approximated as a circle of radius R (see figure). Use Eqs. (6.17) and (6.30) to derive an equation for the stress that develops on the top and bottom face of the cantilever due to bending. If a force of 1 nN is applied to the end of a rectangular cantilever with $L = 400\mu m$, $w_c = 50\mu m$ and $t_c = 2\mu m$, evaluate the stress that develops on the top and bottom faces. Compressive stress is designated as negative, while tensile stress is designated as positive.

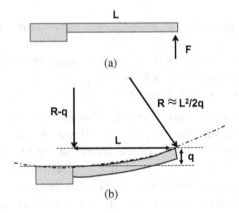

(a)

(b)

7. An interesting property of noise is that it adds in quadrature. Suppose there are say three noise sources that individually produce rms voltages of say e_1, e_2, and e_3. If these noise sources are independent, then the total rms value of the noise (E_{tot}) is given by

$$E_{total} = \sqrt{e_1^2 + e_2^2 + e_3^2}.$$

Suppose one day you reflect a laser beam from a soft cantilever into the photodiode position sensitive detector of your AFM. You happen to notice that the voltage output of your AFM photodiode is noisier than usual. You worry that some electronics has failed since you last used the AFM.

With some thought, you realize that ideally, there should be only three possible sources for the excessive noise: (i) the thermal fluctuations of the cantilever, (ii) intensity fluctuations in the laser beam intensity, or (iii) electronic noise from the photodiode amplifier itself. Since you know how noise adds from different noise sources, you immediately remove the soft cantilever and insert a stiffer one, leaving the laser and photodiode unchanged. You find that the total rms noise voltage from the photodiode decreases. What do you conclude about the origin of the noise?

8. When using sensitive equipment like AFM electronics, it is valuable to have some general awareness about noise generated by operational amplifiers (op amps). The noise specification for an op amp is usually given in units of $nV/\sqrt{Hz}$. In what follows, let's say the typical noise specification for OPAMP Model 123 is about $10\,nV/\sqrt{Hz}$. This noise specification allows an estimate for the input noise to the Model 123 op amp.

   (a) Suppose you have a circuit designed around the Model 123 op amp that is used over a frequency range from 20 Hz to 50 kHz. Using Eq. (6.52), estimate the equivalent input noise to the op amp?

   (b) If the op amp circuit has a gain of 50, what is the output noise of the op amp?

   (c) If you are attempting to measure a $0.2\,\mu V$ signal within the bandwidth of the op amp circuit (i.e., an output voltage of 10 μV), what would you predict for the signal to noise ratio? Will you be able to measure the 0.2 μV signal or must you purchase a less noisy op amp?

9. In dynamic AFM experiments, the cantilever is made to oscillate near its resonance frequency by supplying an AC voltage to a piezo driver located near the cantilever chip. Novice AFM users often "overdrive" the cantilever by using an AC voltage that is too large. One way to detect this situation is to measure the power spectral density (PSD) of the signal from the photodiode when the cantilever driving voltage has a frequency close to the resonant frequency of the cantilever.

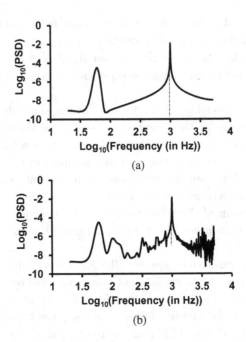

(a)

(b)

Typical PSDs from the photodiode signal are included above. Plot (a) is for a low driving voltage; plot (b) is typical for a higher driving voltage. For clarity, no random noise signal has been included in these two plots. Answer the following questions by referring to the figure provided.

(a) What is the resonance frequency of the cantilever?

(b) Is there any evidence for line voltage contamination of the photodiode signal around 50–60 Hz?

(c) Is the cantilever overdriven at the higher driving voltage? To answer this question, you must compare the harmonic distortions to the main signal.

(d) From the data provided, is it possible to tell whether the 2nd cantilever mode is inadvertently excited at the higher driving voltage? Explain your answer.

Chapter 7

AFM System Components

We are now in a position to discuss the AFM instrument itself. This is a difficult task because each reader of this chapter will likely have access to a different AFM which has its own unique advantages and limitations. Some may be using a home-built instrument, others might be using an instrument that is ten to fifteen years old, and yet others may have just purchased a new instrument which includes all the latest technology. It is therefore difficult to know how best to structure the discussion that follows. The approach we will follow is to give a broad overview of the different subsystems that comprise an AFM and to describe the underlying principles.

The AFM is unique because it is a proximal probe microscope — it probes very local properties. The topographic images obtained are truly 3-dimensional. When used to its fullest advantage, it can provide not only topographic information but also local material properties.

In this chapter we discuss the various components of an AFM with an emphasis on how the various parts are integrated into a system. In general terms, such topics are usually of most interest to mechanical engineering and physics students who are not only interested in equipment design but also in system integration.

7.1 The Generic AFM System

At a very high level, an AFM consists of four major components: (i) an AFM head, (ii) hardware and electronics that adjust the position of a sample in a vibration-free fashion, (iii) a control loop which maintains a specified

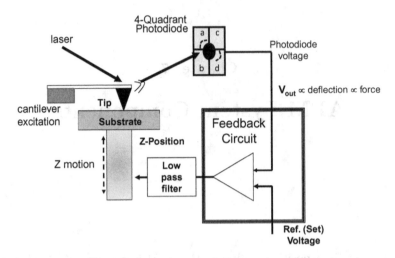

Fig. 7.1   A schematic diagram illustrating the essential components of an atomic force microscope.

quantity (e.g., the force between tip and sample) constant during the course of an experiment, and (iv) a computer and controlling software.

A schematic view of the essential AFM components (i) through (iii) are laid out in Fig. 7.1. This figure shows a tip affixed to a cantilever in close proximity to a substrate. Tip–substrate forces (repulsive when the tip is in contact with the substrate or attractive when the tip is suspended above the substrate) act on the tip and cause the cantilever to deflect, either away from or toward the substrate. The main purpose of the AFM system is to keep the tip–substrate force constant over time. This is done by attaching the substrate to a z-positioner that allows for a fine, vibrationless adjustment of the sample's vertical position. The deflection of the cantilever is monitored by a laser beam that reflects from the back-side of the cantilever onto a segmented photodiode. A voltage from the photodiode, V_{out}, is used to monitor changes in the cantilever position. When V_{out} is compared to a set reference voltage V_{set} (preset by the user), small changes in the cantilever's position can be monitored with high sensitivity. A signal is generated by a feedback circuit that is proportional to the difference between V_{out} and V_{set}. This difference (error) signal is then filtered and sent to the z-positioner. The error signal tells the z-positioner to move up or down in order to make V_{out} equal to V_{set}. The primary function of the low pass filter in Fig. 7.1 is to remove high frequency noise in the signal to the z-positioner. This noise can cause a jitter in the sample's z-position.

Typically, the high frequency cut-off in the low pass filter is set far below internal resonances of the z-positioning stage.

The AFM system outlined in Fig. 7.1 functions like a closed loop mechano-electric controller designed to keep the tip–substrate force constant over time. The deflection of the cantilever essentially monitors the tip–substrate interaction in real time and produces a signal by way of a reflected laser beam. The reflected laser beam strikes the photodiode which generates a time dependent signal proportional to the cantilever's position. This signal is then compared to a constant set signal specified by the user. Based on the difference between these two voltages, a signal is sent to the z-positioner to adjust the substrate position in order to maintain a constant cantilever deflection over time. If the z-positioner can respond quickly enough with sufficiently high accuracy, then nanometer-scale adjustments to the substrate's position can be made without introducing instabilities or uncontrolled oscillations.

In dynamic AFM, the cantilever is forced to oscillate near its resonant frequency to cause a sinusoidal motion of the tip above the substrate. This task, accomplished by means of a separate cantilever excitation assembly, forms an important subsystem of any AFM. The ability to vibrate the cantilever is indicated schematically in Fig. 7.1 by the rectangular element labeled cantilever excitation.

The following sections discuss the operation and implementation of the various subsystems required to make an AFM system operative.

7.2 Transducing the Tip–Substrate Force

Following the discussion in Chapter 3, it should be clear that the tip will, in general, feel an attractive force if it is positioned above a substrate. This attractive force is the sum of long-range collective tip–substrate interactions and much shorter range atom–atom interactions. This force can be measured by attaching the tip to a cantilever and then measuring the deflection of the cantilever due to tip–substrate forces. This approach was proposed in the very first AFM described in 1986 and has persisted to the present. The cantilever acts as a linear spring with a spring constant k determined by the material and dimensions of the cantilever (see Eq. (6.18)). If the spring constant is chosen wisely, the tip–substrate interaction force will be balanced by the cantilever deflection so that a force-balance equilibrium condition is reached. By measuring the cantilever deflection and knowing

Table 7.1   Typical characteristics for three different classes of
cantilevers often used in AFM experiments.

Typical use	k (N/m)	Resonant frequency (kHz)
Non-contact mode	10–100	100–300
Intermittent contact	1–10	20–100
Contact mode	0.1–1	1–50

the spring constant, it is then possible to infer the tip–substrate interaction
force.

There are a wide variety of different AFM cantilevers that can be
purchased and each has its own advantages and disadvantages. Table 7.1
broadly summarizes the spring constants, resonant frequencies, and typical
use for which many of the cantilevers that can be purchased commercially.
Often an AFM cantilever is coated with a thin layer of metal (Al or Au) to
increase its reflectivity.

There have been a variety of methods used to measure the cantilever
deflection versus time, including

1. measurement of the tunneling current from a metallic STM tip positioned on the back-side of the cantilever [binnig86a],
2. laser heterodyne interferometer [martin87],
3. fiber-optic interferometry [rugar89],
4. capacitance changes measured with respect to a fixed plate [göddenhenrich90],
5. piezoresitive cantilevers [tortonese93],
6. focused laser beam bounce [marti88], [meyer88], [alexander89].

By far, the most popular implementation is the laser beam bounce
method. Most commercially available AFMs use a split photodiode to accurately monitor the position of a focused diode laser as the beam spot is
reflected from the backside of a cantilever. As we will see, this approach
allows an accurate and sensitive method to monitor deflections of the
cantilever.

7.3   Detecting Cantilever Deflection

The cantilever motion is monitored by a laser beam reflected from the backside of the cantilever. The situation is sketched in Fig. 7.2, which shows the

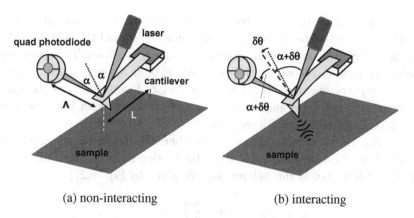

(a) non-interacting   (b) interacting

Fig. 7.2 Illustration of the alignment of the laser, a cantilever (length L) rigidly clamped at one end, and a 4-segment quad photodiode (a distance Λ from the cantilever) for (a) an undeflected (non-interacting) tip and (b) when the cantilever is deflected through an angle $\delta\theta$ due to an attractive tip–sample interaction.

basic geometry. Rarely is the alignment of these components so obvious as pictured since ancillary mirrors and prisms are often used to direct the laser beam onto the cantilever in a confined space.

As shown in Fig. 7.2(a), a laser diode produces a focused light beam which reflects from the backside of a cantilever. Typically, the laser operates at 5 V dc and requires $\sim$50 mA to produce a light beam with a power P_o of $\sim$1 mW and a diameter of a few millimeters. When unfocussed, the light from a laser diode typically forms a beam with an elliptical cross-section. To better define the laser spot on the cantilever, the laser diode is integrally fitted with an external lens assembly (not shown). For a laser with power P_o, the intensity of light across the focal spot, $I(r)$ (W/m^2), is approximated by a Gaussian intensity distribution

$$I(r) = \frac{4P_o}{\pi D^2} e^{-\left(\frac{2r}{D}\right)^2} \tag{7.1}$$

where r is the radial displacement from the center of the focused laser beam and D is a parameter that represents the beam diameter in the focal plane. The diameter D of the focal spot on the backside of the cantilever can vary widely depending on the intrinsic quality of the diode laser and can usually be minimized by a careful optical adjustment of the focusing lens. Values for D typically lie in the range between 5 to 50 microns. Generally speaking, a smaller spot allows a more precise alignment of the beam onto the cantilever.

In Fig. 7.2(a), the situation is shown when the tip is not interacting with the sample. The cantilever is then undeflected and the incident laser beam strikes the cantilever at an incident angle α as shown. The specularly reflected beam bounces from the cantilever, making an angle α with respect to a line perpendicular to the cantilever surface before it strikes the quad photodiode. The total angular deflection of the laser beam from the non-interacting (straight) cantilever is 2α.

In Fig. 7.2(b), the situation is shown where the free end of the cantilever is displaced by a distance q due to an interaction force with the substrate. The cantilever bends through an angle $\delta\theta$ given by Eq. (6.21)

$$\delta\theta = \frac{3}{2}\frac{q}{L} \tag{7.2}$$

where L is the length of the cantilever.

The cantilever's surface normal is rotated by the same angle $\delta\theta$ and as a result, the laser beam now strikes the cantilever at an incident angle of $\alpha + \delta\theta$. For specular reflection, the reflected laser beam bounces from the cantilever at the same angle $\alpha + \delta\theta$. Now, the total angular deflection of the laser beam is $2(\alpha + \delta\theta)$. The important result is that the change in angular deflection of the laser beam is twice the angle of the cantilever's deflection. This causes the reflected laser spot to shift its position on the segmented photodiode as shown schematically in Fig. 7.2(b).

The shift in the reflected laser spot by a distance Δ_q on the photodiode is shown in more detail in Fig. 7.3. Here, for convenience, the photodiode is comprised of only two separate segments top (T) and bottom (B) and each segment produces a voltage proportional to the intensity of the incident light that strikes it.

An operational amplifier connected to the output of the photodiode produces an output voltage proportional to the voltage difference ($V_{\text{diff}} = V_{\text{T}} - V_{\text{B}}$). In this way, a voltage signal related to the relative displacement of the laser spot across the photodiode is produced as indicated schematically in Fig. 7.3(c).

If the position of a circular laser spot is initially positioned to equally illuminate both segments T and B, then a null condition is achieved as indicated in Fig. 7.3(c). If the alignment of the laser beam to achieve the null condition is performed when the tip is far from the sample, then the null condition can be taken to represent the cantilever when no force is acting on the tip. Once the output voltage is nulled, the gain of the operational amplifier can be increased, allowing the detection of a small deflection of

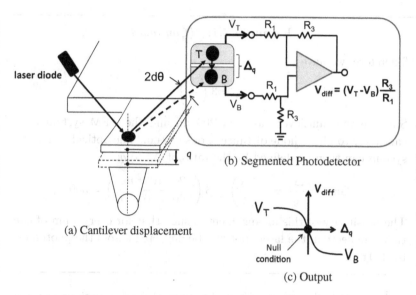

(b) Segmented Photodetector

(a) Cantilever displacement

(c) Output

Fig. 7.3   An illustration of the motion of the reflected laser beam across a bi-segmented photodiode due to cantilever displacement. In (a), the laser spot is displaced by a distance Δ_q when the cantilever displacement is q. In (b), an operational amplifier wired to provide the difference between V_T and V_B. In (c), the variation in the output of the operational amplifier as the laser spot sweeps across the photodiode. When the reflected laser spot is symmetrically placed, no output voltage is measured and a null condition results.

the cantilever due to tip–substrate forces with great sensitivity. For small cantilever displacements q, V_{diff} is linear with spot displacement.

Example 7.1:   Estimate the displacement of the reflected laser spot on the photodiode for the laser beam bounce system shown in Fig. 7.2. Assume the cantilever length is equal to L and the distance between the cantilever and the quad photodiode is equal to Λ.

When the laser beam is deflected through an angle $\delta\theta$, the laser spot will be displaced across the photodiode by a distance Δ_q where

$$\Delta_q \simeq 2\Lambda\delta\theta.$$

For a cantilever displacement q, the angular deflection of the cantilever is given by Eq. (7.2)

$$\delta\theta = \frac{3}{2}\frac{q}{L}.$$

(*Continued*)

Example 7.1:   (*Continued*)

Therefore, we have

$$\Delta_q = 3 \left(\frac{\Lambda}{L}\right) q.$$

Suppose the cantilever has $L = 200\,\mu\text{m}$ and the AFM system has a cantilever to photodiode distance Λ of say 2 cm. The optical gain of the system can then be defined by the ratio of Δ_q to q:

$$\text{Gain} \approx \frac{\Delta_q}{q} = 3\left(\frac{\Lambda}{L}\right) = 3\left(\frac{2 \times 10^{-2}\,m}{200 \times 10^{-6}\,m}\right) = 300.$$

The sensitivity of this arrangement is such that for every 1 nm of cantilever deflection, the laser spot will be deflected across the photodiode by 300 nm.

The photodiode can also be used to correct for fluctuations in the output power of the laser by including a separate circuit (not shown) which produces the sum of the voltage from segments T and B ($V_T + V_B$). With additional summing circuits (not shown), an output voltage can be generated by forming the ratio

$$V_{\text{out}} = \frac{V_T - V_B}{V_T + V_B}. \tag{7.3}$$

Since V_{out} is proportional to V_{diff} (Fig. 7.3) and V_{diff} is proportional to the cantilever angular deflection for small deflections, and since $\delta\theta$ is proportional to the force F (Eq. (6.20)), we can then write to a good approximation that

$$V_{\text{out}} = S_z \cdot F, \tag{7.4}$$

where S_z (units of V/nN) is a calibration factor that must be determined experimentally.

Since the cantilever both deflects (q) and rotates ($\delta\theta$), it is a valid question how these two quantities combine to produce a displacement of the laser spot on the photodiode. As was shown in Chapter 6, both are proportional to the applied force. This question is left to a homework problem, but the answer is that the angular deflection of the cantilever produces by far the largest displacement of the reflected laser spot on the photodiode.

7.4 Detecting Cantilever Torsion

Any twist that the cantilever might experience can be detected using a photodiode assembly as shown schematically in Fig. 7.4. In contrast to detecting the normal deflection of the cantilever, the lateral twist requires a photodiode segmented in a direction perpendicular to that shown in Fig. 7.3.

If the cantilever twists through an angle $d\varphi$, then the reflected laser beam deflects through an angle $2d\varphi$ as indicated. This causes the reflected laser spot to shift its position by an amount Δ_φ on the segmented photodiode as shown schematically in Fig. 7.4(b). The resulting situation is analogous to the discussion already presented for Fig. 7.3.

Now, the photodiode is comprised of two separate segments L and R. An operational amplifier connected to the output of the photodiode produces a signal voltage that is proportional to the difference in the voltages $(V_L - V_R)$. In this way, a signal related to the displacement of the laser spot across the photodiode can be produced as indicated schematically in Fig. 7.4(c).

If the location of the laser spot is initially positioned to equally fall on both segments L and R, then a null condition is achieved as indicated in Fig. 7.4(c). This adjustment can then be taken to represent the condition of the cantilever when no force causes it to twist.

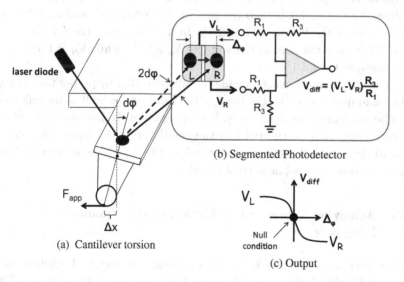

(b) Segmented Photodetector

(a) Cantilever torsion

(c) Output

Fig. 7.4 An illustration of the motion of the reflected laser beam across a bi-segmented photodiode due to cantilever torsion. The laser spot is displaced by a distance Δ_φ when the cantilever is twisted through an angle $d\varphi$.

The photodiode can also be used to correct for fluctuations in the output power of the laser by including a separate circuit (not shown) which produces the sum of the voltage from segments L and $R(V_L + V_R)$. With this voltage sum, a final output voltage can then be generated by forming the ratio

$$V_{\text{out}} = \frac{V_L - V_R}{V_L + V_R}. \tag{7.5}$$

Since V_{out} is related to the cantilever angular deflection, and since $d\varphi$ is proportional to the lateral force F_{app}, we can then write

$$V_{\text{out}} = S_\varphi \cdot F, \tag{7.6}$$

where S_φ is a calibration factor that must be determined experimentally.

For clarity, the discussion in Secs. 7.2 and 7.3 treats the photodiode as a bi-segmented device, but in most AFMs, the photodiode actually contains four active regions (See Fig. 7.1). Four-quadrant (quad) photodiodes are readily available for use in applications related to fast position-sensitive light detectors since they typically have response times under 50 ns.

A quad photodiode consists of four separate photodiodes, with each quadrant separated from the other three by a small gap. Thus segment T in Fig. 7.3 could be subdivided into two equal quadrants a and c, while segment B is split into two quadrants b and d as shown in Fig. 7.1. The voltage V_T in Fig. 7.1 is then the sum of the voltages V_a and V_c, while V_B is the sum of the voltages V_b and V_d. In a similar way, the voltage V_L in Fig. 7.4 is then the sum of the voltages V_a and V_b while V_R is the sum of the voltages V_c and V_d.

The advantage of using a quad photodiode is that by proper use of operational amplifier circuits, the twist of the cantilever as well as its deflection can be monitored simultaneously. It is common in the literature to call the voltage generated by the twist of the cantilever as the lateral channel or lateral signal and the voltage generated by the deflection of the cantilever as the normal channel or normal signal.

7.5 Achieving Vibrationless Motion at the Nanometer Length Scale

When certain crystalline materials experience an external applied force in a well-defined direction, they produce a measureable voltage. This phenomenon, discovered in 1880, is referred to as piezoelectricity. Historically, crystalline solids are the largest class of materials that exhibit the piezoelectric effect. The connection of piezoelectricity to atomic structure is now well known, and only crystalline materials that lack a center of symmetry can be piezoelectric. The underlying fundamental mechanism is an alignment of electric dipole moments in a crystallographic unit cell to produce a net electric polarization. The inverse piezoelectric effect, in which a displacement is produced upon application of a voltage, is of most interest within the context of an AFM. Of considerable interest is that a reversal of the applied voltage reverses the resulting displacement. While initially a scientific curiosity, piezoelectric materials are now widely used to achieve nanometer motion by the simple application of a voltage signal. In this way, no mechanical linkages are required to displace an object.

In the 1950's, it was found that fine piezoelectric powders could be calcined into polycrystalline ceramics, thereby enabling piezoelectric materials with novel shapes (disks, rods, tubes, etc.). It is natural to suspect the spontaneous polarization in each of these fine crystalline grains has a random orientation, producing a net cancellation as a whole with the result that the pieozceramic exhibits no piezoelectric behavior at all. But when a high D.C. voltage is applied to such ceramics in a process known as poling, the direction of the randomly polarized grains are aligned and a piezoelectrically-active ceramic can result. The fraction of the grains that align depends upon the poling voltage, the temperature, and the time the voltage is held on the material. Piezoceramic materials are usually based on Lead Zirconate Titanate (PZT) and Barium Titanate, although current work is focused on developing lead-free piezoceramic materials [rodel09].

Typically, a material is piezoelectric if an atomic dipole moment arises due to an asymmetric shift in the position of atoms in the unit cell of a crystal. The density of dipoles (dipole moment per unit volume) that develops defines a polarization P (units of C/m^2) which can be made to vary with an applied mechanical stress applied across a piezo-active material. Typical values of P lie in the range of 10–$100\,\mu C/cm^2$. The stress reconfigures atoms in the solid, causing a reorientation of internal dipoles. The net result is an equal but opposite charge build up on opposite surfaces of the piezo material. This charge build up is equivalent to a net internal dipole moment which equals the surface charge times the effective distance between the charges (usually equal to the thickness of the piezo bar).

The polarization P physically describes the bound surface charge density by $\sigma_b = \vec{P} \cdot \hat{n}$ ($\hat{n}$ is unit vector perpendicular to surface) or a surface charge $q_b = \vec{P} \cdot \hat{n}\, dA$ that develops on a surface element dA of a polarized material. In principle, P will be an unknown function of position, even if

Piezoelectric Bar

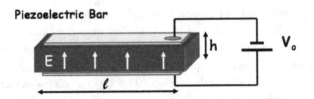

Fig. 7.5 A schematic for a piezoelectric bar. When heated to a temperature slightly below the Curie temperature of the material, an applied electric field E serves to align the crystalline dipoles of the piezo material in a preferred orientation. Before applying the electric field, the crystalline unit cells are cubic in structure and have no net dipole moment. After E is applied, a preferred tetragonal distortion of the unit cells develops. The unit cell distortion remains at room temperature even after the field is removed.

the poling is performed carefully. An applied strain causes excess charge buildup just below the surface (excess charges at the surface tends to be neutralized by adsorbed gas atoms and molecules from the ambient), which in turn produces an electric field within the piezo material. Equivalently, as shown in Fig. 7.5, by applying a voltage to electrodes plated on the piezo material, an electric field will be produced and a mechanical stress will develop that changes the dimensions of the piezoelectric material.

One of the most widely used piezoelectric ceramics is a mixture of lead titanate ($PbTiO_3$) and lead zirconate ($PbZrO_3$). This ceramic, commercially designated as PZT-5A, has a Curie temperature near 350°C. When soldering electrical wires to metallic electrodes plated on the ceramic, temperatures should be held below 300°C for short periods of time in order to prevent degradation of the polarized material. The dielectric constant κ of PZT-5A is about 1700.

A useful configuration is shown in Fig. 7.5 where a piezoelectric bar of length ℓ and thickness h is sketched. The exterior surface of the bar is plated with two conducting electrodes (Au, Ni, Ag, Cu) as shown. The polarization vector P is pre-aligned along the direction defined by the two electrodes. Such an alignment of P is accomplished by heating the piezoelectric in the presence of a polarizing applied voltage, and then slowly cooling the crystal while maintaining the applied voltage. After poling, if the piezo material is subjected to high temperatures or thermal shock, the alignment of P can be disturbed and the performance of the piezoelectric material can degrade.

Upon application of an applied voltage V_o, the fractional length of the bar changes by

$$\frac{\Delta \ell}{\ell} = d_{31} \frac{V_o}{h} \qquad (7.7)$$

where d_{31} (C/N or m/V) is the appropriate transverse strain coefficient for the material in use. Although d_{31} is usually presented as a constant factor, its value does vary with temperature, pressure, electric field, form factor, and mechanical/electrical boundary conditions. The sign of $\Delta\ell$ — either expansion or contraction — depends on the polarity of the applied voltage. Typically $|d_{31}|$ lies in the range between 10 and 100 pm/V. For a piezo bar with a 1 cm length that is 1 mm thick. Assuming $|d_{31}| = 50$ pm/V, it is easy to show that the bar will changes its dimension by $\Delta\ell = 0.5$ nm for every volt applied. Since it is usually an easy matter to adjust a dc voltage by ± 10 mV, systematic length changes of ~ 50 pm or less can be achieved.

Example 7.2: Piezoelectricity can generate an internal force on surrounding structures whenever a piezo bar is rigidly confined, causing unintended internal stresses to develop. Sometimes the stresses can be so high the piezo bar will fracture.

Suppose 10 V is applied to a piezoelectric bar that is constrained to fit between two rigid supports. How much stress does the piezoelectric bar generate? Assume the bar is 1 cm long with a 1 mm × 1 mm cross-sectional area. The piezo coefficient d_{31} is 50 pm/V and Young's modulus of the piezo ceramic bar is 70 GPa.

If the bar were free to expand, it would change its length according to Eq. (7.7):

$$\frac{\Delta\ell}{\ell} = d_{31}\frac{V_o}{h} \Rightarrow \Delta\ell = d_{31}\frac{V_o}{h}\ell = \left(\frac{50\,\text{pm}}{\text{V}}\right)\left(\frac{10\text{V}}{0.001\text{m}}\right)(0.01\text{m}) = 5\,\text{nm}.$$

Since the piezo bar is rigidly confined, the support structure must exert a compressive force to maintain the original 1 cm length. The strain that develops in the piezo during this compression must therefore equal $\Delta\ell/\ell$ as calculated above. Since Young's modulus is the ratio of stress to strain, we have

$$E = \frac{\text{stress}}{\text{strain}}$$

$$\text{stress} = \frac{\Delta\ell}{\ell}E = \frac{5 \times 10^{-9}\,\text{m}}{0.01\,\text{m}} \cdot 70\,\text{GPa} = 35\,\text{kPa}.$$

For sufficiently high applied voltages, this stress can cause unintended distortions in the geometry of the support structure.

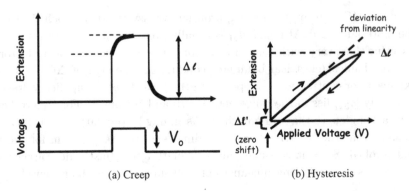

(a) Creep   (b) Hysteresis

Fig. 7.6   Schematic diagrams illustrating (a) piezo-creep (emphasized by bold line segments) and (b) piezo-hysteresis. In (a), a sudden increase in the voltage applied to a piezo material produces a gradual change in the extension $\Delta\ell$ of the material. In (b), a linear sweep up and down of the applied voltage produces an extension $\Delta\ell$ that exhibits hysteretic behavior as depicted by $\Delta\ell'$. Experience indicates these unwanted non-linearities tend to become more important as the value of d_{31} increases or as the applied voltage increases.

Ideally, the deformation induced by the applied voltage is completely reversible, but this is hardly ever the case. Defects in the piezo material cause trapped charge to accumulate over time. In addition, a slow reorientation of domains within the material occurs. These effects contribute to piezo creep and hysteresis as pictured in Fig. 7.6.

Piezo creep occurs when the extension of the piezo material cannot instantaneously follow the change in applied voltage. Piezo hysteresis refers to a non-linear effect in which the extension of the piezo bar is not linear with applied voltage. Usually, this effect is most noticeable when the applied voltage swing is large. In addition, when the voltage is returned to its original value, the piezo bar will not immediately return to its original dimension. Both these effects can be somewhat mitigated by "exercising" the piezo, that is by applying a repetitive voltage waveform that causes multiple expansion and contractions to occur.

Apart from high operating temperatures, creep, and hysteresis, it is also known that the performance of piezoelectric materials degrade with age as the polarization of individual grains start to randomize to a lower energy configuration or when high operating electric fields are required.

To achieve motion in X, Y and Z, three piezo bars could be glued together in an orthogonal fashion. Alternatively, a four-quadrant piezo tube (a thin-walled, cylindrical tube made from a piezo ceramic material) is commonly employed [binnig86b].

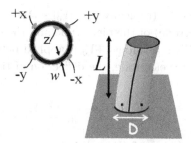

Fig. 7.7 A schematic of a piezoelectric tube scanner of length L and outer diameter D. The thickness of the tube is w. The tube scanner has four separate outer electrodes $(+x, -x, +y, -y)$ and a continuous inner electrode (z). By applying voltages to each of the four outer electrodes, the top of the piezo tube can be made to bend in the $+x, -x, +y, -y$ directions. Typically, a piezo tube used for AFM applications will generate a 25–50 nm displacement for every volt applied.

Typically, a piezo tube has one continuous inner electrode plated around the entire inside surface of the hollow cylinder (z) and four electrodes $(+x, -x, +y, -y)$ placed symmetrically around the exterior surface. A schematic is given in Fig. 7.7. The tube has a diameter D, a length L and a thickness w and contracts both radially and longitudinally depending on the applied voltage. One edge of the tube (bottom edge in Fig. 7.7) is rigidly held in place, while the other is free to deform under the strains produced by the applied voltage. A limitation of a tube scanner is the possibility of unequal quadrants due either to imbalanced geometrical placement of the electrodes or variability in electrical poling between the inner and each of the four outer quadrants. Quadrants with unequal properties lead to bizarre scanning effects.

A simple way to check the integrity of the piezo tube is to first calculate the expected capacitance between the inner and outer four electrodes. This capacitance can be estimated by using the formula for the capacitance of concentric tubes of radii R_{outer} and R_{inner}, with a length L (all in m) and filled with a material having a dielectric constant κ

$$C = \frac{2\pi\kappa\varepsilon_o}{\ell n\left(\frac{R_{outer}}{R_{inner}}\right)L} \qquad (7.8)$$

Here, the dielectric permittivity of vacuum (ε_o) has a value of $8.85\,\text{pF/m}$. Assuming roughly equal area electrodes, the capacitance of each quadrant should have about the same value and should equal 1/4 of the value calculated using Eq. (7.8) above.

A scanning voltage V_o (typically $0 < |V_o| < 100\,\mathrm{V}$) is applied between the inner wall (z-electrode) of the piezo tube and the outer four electrodes ($+x$, $-x$, $+y$, $-y$). When a voltage $+V_o$ is applied between the continuous inner and any one outer electrode, the deflection of the piezo tube will be given by [chen92]

$$\Delta x \simeq \Delta y = \sqrt{2}\, d_{31} \frac{L^2}{\pi D} \frac{V_o}{w}. \tag{7.9}$$

The elongation/contraction of the tube along the z-axis is obtained by applying a voltage V_o between the inner wall (electrical ground) and *all* outer electrodes. The change in z is then given by

$$\Delta z = d_{31} L \frac{V_o}{w}. \tag{7.10}$$

In the symmetric voltage mode of operation, the drive voltage is bipolar (i.e., $+V_o$ is applied to the $+x$ electrode and $-V_o$ is applied to the $-x$ electrode) and the resulting deflections in x and y are twice as large as given by Eq. (7.9). Under these conditions, one quadrant extends the other contracts and the free end of the tube bends due to the combined stresses. Finite element calculations for tube deflection provide a more realistic assessment of displacement vs. applied voltage than the analytical formulas above [carr88].

While small and inexpensive, tube scanners do have some well recognized limitations. First, they have a relatively low resonance frequency, on the order of $1\,\mathrm{kHz}$. Low resonant frequencies imply low scanning speeds. Piezotubes also exhibit creep and hysteresis behavior discussed in Fig. 7.6. These limitations conspire to produce a noticeable shift between forward and backward images. Lastly, since positions in x, y and z are controlled by one tube, issues arise about the orthogonality of the scans. Ideally, the changes Δx, Δy, and Δz should be independent of each other, and while this is approximately true, there can be significant coupling between these quantities as the driving voltage becomes large.

Recent trends in AFM technology now make use of newly developed flexure stages which offer many advantages over piezos. Flexure stages have zero mechanical friction and the flexure hinges (formed from precision slots cut directly into the stage) restrict the motion of each axis to a single direction, effectively producing a pure, 1-dimensional translation. In addition, feedback control is implemented using capacitive, piezoresistive or inductive sensors to allow reproducible positioning that can be calibrated using optical interferometric standards. Maximum scan ranges vary from $\sim 0.5\,\mu\mathrm{m}$ to

~500 μm, depending on the actuator and design of the flexure stage. The scan speed is restricted by the resonant frequency of the flexure stage, which typically lies in the range from 100's of Hz to 1 KHz. There is minimal out-of-plane curvature over the entire scan range and out-of-plane movement of the stage over the entire scanning range is typically only a few nanometers. With the use of flexure stages in AFMs, linear positioning becomes independent of scan rate, scan range, and scanner offset, providing accurate and repeatable positioning capabilities.

7.6 Exciting the Cantilever

A fundamental requirement to perform dynamic AFM measurements is to induce a sinusoidal motion of the cantilever at a desired frequency. There are many ways to accomplish this task and each method has its advantages and disadvantages. The four main techniques are (a) direct piezoelectric excitation, (b) magnetic excitation, (c) Lorentz force excitation and (d) photothermal excitation. Each technique is illustrated in Fig. 7.8.

Direct piezoelectric excitation (also called base excitation or acoustic excitation) is illustrated in Fig. 7.8(a). This mode of excitation requires that the cantilever base be firmly attached to a small dither piezo which is subjected to an applied AC sinusoidal voltage. The resulting sinusoidal change in shape of the dither piezo drives the cantilever and causes it to oscillate at a frequency determined by the drive voltage. The major disadvantage is the mechanical motion of the base often couples to the surrounding environment, especially when the cantilever is submerged in a liquid, causing unwanted resonances that are often difficult to untangle from that of the cantilever.

To eliminate this difficulty, three other techniques for direct cantilever excitation (no base motion) have been developed. In the magnetic excitation method shown in Fig. 7.8(b), the cantilever is coated with a thin film of ferromagnetic material which is then polarized with a small bar magnet. When the cantilever is brought into close proximity to an oscillating magnetic field, the magnetic properties of the thin film then produce a force directly on the cantilever. The magnetic field can be produced by a small coil of wire located near the cantilever.

An alternative approach is to produce a cantilever with a continuous, electrically conducting loop as shown schematically in Fig. 7.8(c). The loop supports a current $i(t)$ which is driven at a fixed frequency determined

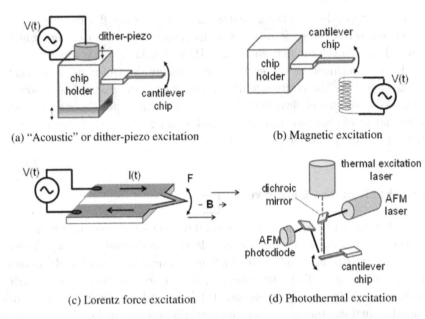

(a) "Acoustic" or dither-piezo excitation   (b) Magnetic excitation

(c) Lorentz force excitation   (d) Photothermal excitation

Fig. 7.8   Four techniques used to excite a cantilever into sinusoidal motion. In (a), base excitation of entire cantilever chip by a small dither piezo. In (b), magnetic excitation by an AC magnetic field produced by a nearby coil. The time varying B field interacts with a magnetized thin film evaporated onto the cantilever. In (c), Lorentz-force excitation due to an AC current passing through a loop formed by the cantilever. This current interacts with a constant applied magnetic field B and produces a Lorentz force causing the cantilever to oscillate. In (d), photothermal excitation caused by the adsorption of a modulated laser beam from a thermal excitation laser. The modulated light is adsorbed by a thin film placed on the cantilever. A periodic heating of the cantilever produces a periodic deflection when steady state conditions are reached.

by an AC voltage source. When the loop is positioned in a dc magnetic field of strength B, the cantilever will experience a time dependent force $\vec{F}(t)$ obtained by integrating the Lorentz force around the closed loop as indicated in Eq. (7.11).

$$\vec{F}(t) = i(t) \oint d\vec{\ell} \times \vec{B}. \qquad (7.11)$$

A photothermal technique, illustrated in Fig. 7.8(d), sinusoidally drives the cantilever by illuminating the cantilever with a chopped laser beam of wavelength λ [kiracofe11]. One advantage of this technique is that the cantilever can be driven at high frequencies determined by the modulation frequency of a laser beam rather than the response of a dither piezo. In the photothermal excitation scheme, the cantilever must be coated with a thin adsorbing layer that absorbs strongly at λ. The light adsorbed by the thin film produces a time dependent temperature variation, causing the cantilever to flex in a sinusoidal manner.

The advantage of the three direct excitation schemes illustrated in Figs. 7.8(b–d) is that only the cantilever oscillates. The disadvantage is that custom cantilevers or specialized cantilever coatings are required. As will be discussed in Part II of these lecture notes, the proper choice of cantilever excitation in dynamic AFM has important consequences.

7.7 The Feedback Loop

The feedback loop that appears schematically in Fig. 7.1 is the heart of the control system for an AFM and it forms a universal and vital component for every AFM in use today. The proper setting of the feedback enables reliable AFM operation. The main purpose of the feedback loop is to controllably modify the substrate position so as to keep some important dynamical variable of the system constant. The dynamical variable could be the contact force, the amplitude of cantilever oscillation, or the cantilever frequency, depending on the mode of operation. The underlying function of the feedback loop is given schematically in Fig. 7.9.

Here, a control signal represented by $S(t)$ is used to make a change $\Delta z(t)$ in the substrate position. Although $S(t)$ may vary with time, more often than not it represents a constant voltage that will be compared to

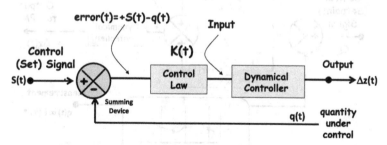

Fig. 7.9 A sketch of a generic feedback loop often used to control the performance of mechanical equipment. An output signal $\Delta z(t)$ from a dynamical controller is sent to the z-piezo of an AFM while the AFM tip is scanning across a substrate. At the same time, the quantity under control $q(t)$ is compared to a set control signal $S(t)$ which is preset by the user. Normally $S(t)$ is constant, independent of time. The control law $K(t)$ produces an input signal to the dynamical controller that minimizes the difference between $S(t)$ and $q(t)$.

the output voltage of the photodiode. The feedback loop contains a summing device which calculates the difference between $S(t)$ and the dynamical variable $q(t)$ to form an error signal given by

$$\text{error}(t) = S(t) - q(t). \tag{7.12}$$

This error signal is used as an input to a control law $K(t)$ which represents an algorithm for minimizing the error signal. $K(t)$ has user controlled settings which allow real time adjustment of the feedback loop's performance. The output of $K(t)$ is used as input to a dynamical controller which produces a voltage that adjusts the experimental quantity under control (tip force, amplitude of the cantilever oscillation, etc.) here specified simply as $q(t)$. When adjusted properly, the error signal is maintained close to zero even though $q(t)$ may vary with time. A proper adjustment of the control loop is critical for the reliable performance of an AFM.

An example of a control law $K(t)$ often used in AFM applications is illustrated in more detail in Fig. 7.10. Here it is assumed that the deflection of the cantilever $q(t)$ is monitored as a function of time, then digitized at equally spaced discrete times $\ldots, t_{n-2}, t_{n-1}, t_n$ and saved for further use. The desired setpoint deflection is preset to some value Q_o. The control law is specified by two terms which attempt to minimize the error at time t_{n+1}.

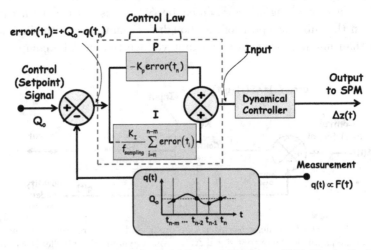

Fig. 7.10   A schematic diagram of a typical control law for an AFM. The feedback loop contains both a proportional (P) and integral (I) controller that attempts to minimize the error signal as a function of time. The proportional controller makes adjustments based on the error in real time while the integral controller has a memory that 'remembers' the error from the recent past.

One term P, with a weight K_P, generates a signal that is **proportional** to the error at time t_n

$$P = -K_P \text{ error } (t_n). \tag{7.13}$$

The second term, I, with a weight K_I, generates a signal that is proportional to the error **integral** back in time from t_n to t_{n-m} with m adjustable

$$I = -\frac{K_I}{f_{\text{sampling}}} \sum_{i=n}^{n-m} \text{ error } (t_i). \tag{7.14}$$

In Eq. (7.14), the sampling frequency is defined as

$$f_{\text{sampling}} = \frac{1}{t_n - t_{n-1}}. \tag{7.15}$$

If the error happens to be zero at t_n, then no proportional correction need be made, but an integral correction may result based on the accumulated past history of $q(t)$. The control law $K(t)$ is the sum of Eqs. (7.13) and (7.14).

In effect, the proportional controller (P) allows the probe to better follow smaller, high frequency features by adjusting K_P. The backward integrator controller (I) allows the probe to follow low frequency features by adjusting K_I. The sampling frequency f_{sampling} is usually determined by the sampling frequency of the analog to digital converter and should be greater than ~10 times the frequency of the fastest variation in $q(t)$ that is being measured in a particular experiment.

Every feedback loop is limited by its inherent bandwidth, which means that the loop simply cannot make adjustments fast enough for reliable operation if $q(t)$ changes rapidly with time. With current technology, the typical bandwidth of an AFM feedback loop is on the order of 10 kHz, and this bandwidth implies that the fastest changes in $q(t)$ that can be accurately tracked should be roughly 1/10 of the bandwidth (~0.1 × 10 kHz = 1 kHz). Since the frequency that any feature in topography appears to the feedback circuit is related to the scanning speed, it follows that the scanning speed must be adjusted to allow a data point approximately every 0.001 s to match the 1kHz criterion above. For an image with 256 × 256 data points, the required time for fastest image acquisition is about

$$\text{fastest acquisition time} \approx \frac{0.001 \, s}{\text{pixel}} \times (256)^2 \, \text{pixels} \approx 65 \, \text{s}.$$

Typically, on flat samples, AFMs provide 256 × 256 pixel images in a few minutes, about 2-4 times slower than the 1 minute fast-time estimated above.

If K_P and K_I are too large, the system will overreact to small errors and become unstable. If K_P and K_I are too small, then the system will react too slowly to errors and long-term drift will set in. While it is difficult to specify with certainty how to tune the feedback parameters, every AFM user should have in mind tuning rule guidelines based on experience. Unfortunately, these guidelines may not be appropriate for a particular sample and you might not even realize it unless great care is exercised. How to efficiently choose K_P and K_I requires a more detailed discussion. The proper setting of the feedback loop will be discussed in more detail in Chapter 9.

Experience shows that setting feedback parameters is a method of successive approximations and real-time tuning of AFM feedback loops is required. Usually, the equipment manufacturer will provide some helpful guidelines. Each AFM system has a range of K_P and K_I that maintain a stable feedback. Careful AFM work, when published, requires at least a comment on how the feedback parameters K_P and K_I were systematically optimized.

7.8   Electronic Control

Scanning commands, monitoring of the output voltage from the photodiode, feedback circuit control, etc. are all controlled by a computer executing a real time data acquisition program from a personal computer. Usually, after the parameters for an AFM experiment are set, real time control of the AFM is passed to a digital signal processor (DSP) imbedded in a computer. The DSP is a specialized micro-processor that is optimized to rapidly perform a number of sequential mathematical operations on a data series.

A reliable AFM requires a stable operating system that is intuitive and easy to use. In addition to hardware control, the computer must also support extensive data analysis capabilities even while an AFM experiment is underway. Every commercially available AFM provides a control system that meets these requirements in a myriad of different ways. Those systems that are more integrated and require less add-ons are desirable because they allow a wider range of experiments as the interests of the user group evolve.

Figure 7.11 provides a high-level, generic overview of the electronic control system required for any AFM system. While the details are different from system to system, the overview provided in Fig. 7.11 summarizes the

Important System (Electronic) Signals

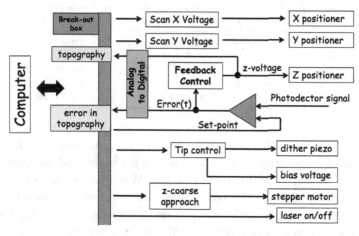

Fig. 7.11 It is difficult to draw a schematic diagram illustrating the control electronics for all AFMs. The above diagram attempts to show those signals which are common to the operation of most AFMs. A breakout box is illustrated that allows access to many of these signals. Usually, the computer contains a digital signal processing board for the dedicated control of the AFM.

essential components of any AFM. Since modern control systems are almost entirely digital in nature, the analog-to-digital converter shown in Fig. 7.11 is often not required.

7.9 Minimizing Thermal Drift

Since the tip is placed ∼1 nm from a substrate and since the X and Y scanners may be required to execute scans of ∼10 nm or less, it is important to mitigate uncontrolled thermal expansion when performing AFM experiments. At issue are uncontrollable temperature gradients that may develop throughout the AFM head. Temperature gradients of a few Centigrade degrees easily develop throughout the system due to heat from electronics, the laser illumination, sample handling and mounting, and light from external illumination sources.

It is well known that a change in temperature leads to thermal expansion or contraction of any material. The change in the length of a bar ΔL which has a length L_o for a given change in temperature ΔT is given by

$$\Delta L = \alpha L_o \Delta T \tag{7.16}$$

Table 7.2   Thermal expansion coefficients of selected materials near room temperature.

Material	$\alpha(°C^{-1})$	Material	$\alpha(°C^{-1})$
Al	23×10^{-6}	Zerodur	$\sim 0.2 \times 10^{-7}$
Steel	$\sim 12 \times 10^{-6}$	Teflon $(C_2F_4)_n$	135×10^{-6}
Au	14×10^{-6}	Alumina (Al_2O_3)	5×10^{-6}
Glass	$3 - 8 \times 10^{-6}$	Si	3×10^{-6}
Invar	1.2×10^{-6}	Quartz	$0.8 - 1.4 \times 10^{-6}$

where $\alpha(°C^{-1})$ is the linear coefficient of thermal expansion (CTE), a material dependent quantity. Table 7.2 provides a list of thermal expansion coefficients for a variety of different materials. When performing experiments where high stability is required, unless care is taken to match the thermal expansion of parts made from different materials, the differing thermal expansion will cause a multitude of different instabilities.

If the temperature varies with time, so will the length according to

$$\frac{dL}{dt} \simeq \alpha L_o \frac{dT}{dt}. \tag{7.17}$$

Example 7.3:   A metal bar 1 cm long used to support a sample in an AFM is subjected to a temperature drift of $1°C/\text{min}$. What is the rate of change of the bar's length?

For metals, typically $\alpha \approx 20 \times 10^{-6} °C^{-1}$. Equation (7.17) implies

$$\frac{dL}{dt} \simeq \alpha L_o \frac{dT}{dt}$$

$$= (20 \times 10^{-6} °C^{-1})(1\,\text{cm}) \left(\frac{1°C}{\text{min}}\right)$$

$$= 20 \times 10^{-6}\,\text{cm/min} \times \frac{1 \times 10^7 \text{nm}}{\text{cm}} = 200\,\text{nm/min}.$$

Mounting a sample to such a metal support and attempting to obtain stable and reproducible data at a length scale less than or comparable to 200 nm will clearly be a futile exercise until the temperature drift stabilizes.

An equally important consideration is how long it will take to dissipate a temperature gradient across a given component. The answer clearly depends on the rate at which the component will reach thermal equilibrium. For this reason, the thermal conductivity of various materials also becomes important. As always, there is a tradeoff, since thermally stable materials with low CTE tend to have low thermal conductivities. Therefore, they require a longer time to reach thermal equilibrium than expansive materials with high CTE, which tend to have high thermal conductivities.

For AFM applications, the simplest strategy to overcome uncontrollable drift effects requires time. It is advisable to load a sample, align the laser beam on the backside of the cantilever, and wait for thermal gradients to dissipate. This often requires waiting for an hour or more before serious data acquisition can begin. For critical experiments at nanometer length scales, a sample and cantilever are mounted, the laser beam is aligned, and the entire AFM system is often left energized overnight with the tip scanning a few microns above the sample.

Another source of thermal drift worth mentioning is related to the reflective coating often added to the backside of the AFM cantilever to increase the light reflected into the photodiode. Since the reflective coating is made from a different material than the cantilever itself, a compound cantilever is created with two separate layers. This compound cantilever is almost always under some residual stress due to the deposition of the reflective material, causing the cantilever to have a slight bend either up or down, depending on the metal film deposition conditions. When illuminated by the laser beam, the cantilever heats up and temperature gradients develop, causing large uncontrollable drifts due to internal stresses in the film evaporated on the cantilever until thermal equilibrium is restored.

7.10 Reducing Unwanted Floor Vibrations

The placement of an AFM system within a lab is often critical for its ultimate performance. Floor vibrations can be a major source of uncontrolled noise as building motors (repetitive vibrations), elevators, and local foot traffic (sudden shocks) produce floor vibrations that can couple into the AFM system. The amplitude of floor vibrations can approach $1\,\mu$m at a characteristic frequency determined by the source of vibration. It follows that the best location for an AFM is in a quiet part of a building, preferably at ground level. If located on an upper floor of a building, the AFM system is best positioned near a structural support column in the wall rather than in the middle of a laboratory room with a suspended floor.

If possible, the spectrum of the floor vibrations should be measured with a vibrometer. Vibrations are usually specified by either acceleration, velocity or displacement measured over 1/3-octave-band frequencies ranging between 1 Hz and ~100 Hz[weaver09]. For frequencies higher than ~100 Hz, the origin of the offending vibration is usually considered to be acoustic.

To understand the vibration literature requires some knowledge of standard definitions. As an example, the audio spectrum from ~20 Hz to ~20 kHz is often divided into ~31 frequency intervals or bands called 1/3-octave bands. If the 19th 1/3-octave band center frequency is defined to be $f_{19}^{\text{center}} \equiv 1000\,\text{Hz}$, then all lower octave band center frequencies can be defined relative to each other using the formula

$$f_{n-1}^{\text{center}} = \frac{f_n^{\text{center}}}{2^{1/3}}. \tag{7.18}$$

For each center frequency defined by Eq. (7.18), bands can be set which span the frequencies between low and high limits respectively defined by

$$f_n^{\text{low}} = \frac{f_n^{\text{center}}}{2^{1/6}} \quad \text{and} \quad f_n^{\text{high}} = 2^{1/6} f_n^{\text{center}}. \tag{7.19}$$

Following this definition produces a constant fractional bandwidth of 23.2% for each 1/3-octave-band. Because the standard 1/3-octave-band frequencies used to characterize vibrations are not usually tabulated, they are listed for reference in Table 7.3. Since vibrations greater than 100 Hz are considered to be acoustic in origin, they are not relevant to this discussion.

Because the vibrations are assumed to be sinusoidal in nature, it is easy to derive analytical formulas for acceleration (derivative of velocity

Table 7.3   Defining the 1/3 octave band frequencies for vibrations less than 100 Hz.

Lower Band Limit (Hz)	Center Frequency (Hz)	Upper Band Limit (Hz)
14.1	16	17.8
17.8	20	22.4
22.4	25	28.2
28.2	31.5	35.5
35.5	40	44.7
44.7	50	56.2
56.2	63	70.8
70.8	80	89.1
89.1	100	112

and multiplication by ω) or displacement (integral of velocity and division by ω) if, say, the velocity of the slab is measured. Thus an estimate for acceleration, velocity or displacement can be had from a measurement of any one of the three.

Vibration criteria (VC) useful for the characterization of lab space are designated by the dependence of velocity, displacement, or acceleration vs. frequency. In this way, the industry standard called "vibration criterion E" (VC-E) is often used to qualify sensitive laboratory settings. VC-E specifies a constant floor velocity of 3 μm/sec for frequencies $8\,\mathrm{Hz} < f < 100\,\mathrm{Hz}$, and a constant acceleration of 150 μm/sec^2 at frequencies below 8 Hz. A well designed laboratory floor for AFM use should meet the NIST-A or NIST-A1 vibration standards. NIST-A1 is particularly stringent and requires an RMS velocity of 3 μm/sec for frequencies less than 4 Hz, and a 0.75μm/sec velocity for frequencies $4\,\mathrm{Hz} < f < 100\,\mathrm{Hz}$.

Many AFMs require a vibration isolation table to reduce floor vibrations. The table should have a large mass and soft spring to produce a low natural resonance frequency. Typically, the table mass is $\sim$100 kg and the pneumatic air legs have a spring constant of order 5 N/m, producing a natural (horizontal) resonant frequency of $\sim$1 Hz with a Q of about 2–3. A careful design of the parts in the AFM head will reduce the need for isolation. The guiding principle is to increase the internal resonance of AFM components by making them small and lightweight, thus pushing their internal resonance frequency well above the 100 Hz range. In this way, floor vibrations will not strongly couple to sensitive components located in the AFM head.

Very often, internal resonances can be identified by directing an audio speaker at the head and slowly varying the input frequency to the speaker. A large, reproducible increase in the noise level of the photodiode output around a well-defined frequency f is usually a good indicator of a structural part in the AFM head with an unwanted resonance frequency at f.

The general principle of isolation from floor vibrations is illustrated schematically in Fig. 7.12. The floor is considered to have a sinusoidal vibration of amplitude Δz_o at some frequency $f < 100\,\mathrm{Hz}$, as shown. The AFM with mass m in Fig. 7.12(a) is suspended or supported by a spring system that attenuates the floor vibration by a factor α ($\alpha < 1$). Figure 7.12(b) illustrates the advantage of a multiple N-stage isolation system (here each stage is considered identical), to produce an isolation factor of α^N, assuming no additional resonant frequencies are introduced by the added isolation systems.

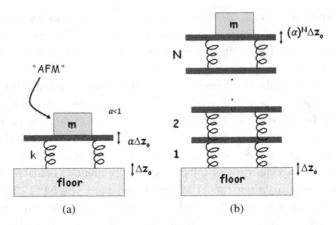

Fig. 7.12   The principle of vibration isolation relies on an attenuation of the vibration of a sample plate by a factor α, where $\alpha < 1$. If one stage of isolation (a) is good, N stages of isolation (b) should be better. In real systems, the attenuation varies with frequency.

7.11   Chapter Seven Summary

This chapter presented a functional review of many of the sub-components common to all AFMs. A generic AFM usually consists of four main components:

(i) An AFM head. Typically, the head contains a cantilever holder, a small laser, and a position sensitive photodiode. From these components, it is possible to detect vertical and lateral cantilever deflections as a tip, attached to the cantilever, interacts with a sample.

(ii) Electronics required for sample scanning. Usually the scanning electronics consists of high-voltage (HV) amplifiers that produce a noise-free signal of a few hundred volts from a computer-generated signal of a few volts. The HV signal is required to move a piezoelectric scanner in a vibrationless fashion over distances ranging from nanometers to micrometers.

(iii) A control loop. Feedback loops are often implemented using a dedicated computer equipped with a DSP. The DSP performs most if not all of the signal processing and image analysis calculations involved in the real-time operation of an AFM.

(iv) Software. All AFM systems require an integrated, custom-designed software package. Raw AFM image files, once obtained, can later be processed using a number of free-ware image-processing programs widely available on the net.

In addition to a discussion of the role and function of these major components comprising an AFM system, considerations related to environmental conditions were also presented. Issues like thermal drift, floor vibrations, etc. must be addressed to insure the successful operation of an AFM.

7.12 Further Reading

Chapter 7 References:

[alexander89]   S. Alexander, L. Hellemans, O. Marti, J. Schneir, V. Elings, P. K. Hansma, Matt Longmire, and J. Gurley, "An atomic-resolution atomic-force microscope implemented using an optical lever", J. App. Phys. **65**, 164–67 (1989).

[binnig86a]   G. Binnig, C.F. Quate, C. Gerber, "Atomic Force Microscope", Phys. Rev. Lett. **56**, 930–33 (1986).

[binnig86b]   G. Binnig and D.P.E. Smith, "Single-tube three-dimensional scanner for scanning tunneling microscopy", Rev. Sci. Instrum. **57**, 1688–89 (1986).

[carr88]   R.G. Carr, "Finite element analysis of PZT tube scanner motion for scanning tunneling microscopy", J. Microsc. **152**, 379–385 (1988).

[chen92]   C. Julian Chen, "Electromechanical deflections of piezoelectric tubes with quartered electrodes", Appl. Phys. Lett. **60**, 132–34 (1992).

[göddenhenrich90]   T. Göddenhenrich, H. Lemke, U. Hartmann and C. Heiden, "Force microscope with capacitive displacement detection", J. Vac. Sci. Technol. A **8**, 383–87 (1990).

[kiracofe11]   D. Kiracofe, K. Kobayashi, A. Labuda, A. Raman and H. Yamada, "High efficiency laser photothermal excitation of microcantilever vibrations in air and liquids" Rev. Sci. Instrum. **82**, 013702–07 (2011).

[marti88]   Marti *et al.*, "Scanning probe microscopy of biological samples and other surfaces", J. Microscopy **152**, 803–09 (1988).

[martin87]   Y. Martin, C. C. Williams, and H. K. Wickramasinghe, "Atomic force microscope-force mapping and profiling on a sub 100 Å scale", J. App. Phys. **95**, 4723–29 (1987).

[meyer88]   G. Meyer and N. Amer, "Novel optical approach to atomic force microscopy", Appl. Phys. Lett. **53**, 1045–47 (1988).

[rodel09]   J. Rödel, W. Jo, K.T.P. Seifert, E-M. Anton, T. Granzow, and D. Damjanovic, "Perspectives on the Development of Lead-free Piezoceramics", J. Am. Ceram. Soc. **92**, 1153–77 (2009).

[rugar89]   D. Rugar, H. J. Mamin, and P. Guethner, "Improved fiber-
                 optic interferometer for atomic force microscopy", App. Phys.
                 Lett. **55**, 2588–90 (1989).
[tortonese93]   M. Tortonese, R. C. Barrett and C. F. Quate, "Atomic res-
                 olution with an atomic force microscope using piezoresistive
                 detection", Appl. Phys. Lett. *62* , 834–36 (1993).
[weaver09]   J. Weaver, M. Voorhis, and R. Reifenberger, "Nanometrology
                 Room Design: The Performance and Characterization of the
                 Kevin G. Hall High-Accuracy Laboratory", J. Inst. of Envi-
                 ron. Sci. and Techn. (IEST) **52**, 1–12 (2009).

Further reading:

[sarid91] D. Sarid and V. Elings, "Review of scanning force microscopy", J. Vac.
           Sci. Technol. **B9**, 431–7 (1991).
[hicks97] T.R. Hicks and P.D. Atherton, *The NanoPositioning Book*, Penton
           Press, London (1997).

7.13 Problems

1. In addition to changes in the physical dimension of a structural component due to a change in temperature, electronic components also change
   their value as they heat up. This can have important consequences since
   resistors are often used as voltage dividers and to set gains in operational
   amplifiers while capacitors find use in filters. Changes in resistance and
   capacitance with temperature can cause a drift in the gain of an operational amplifier or a shift in the cut-off frequency of an electronic filter.
   Such changes can affect the stability of AFM electronics.

   A measure of these drifts is given by the temperature coefficient of
   resistance (TCR) and the temperature coefficient of capacitance (TCC).
   These two quantities are defined by

   $$\text{TCR} = \frac{R(T_2) - R(T_1)}{R(T_1)} \cdot \frac{1}{T_2 - T_1}, \quad \text{TCC} = \frac{C(T_2) - C(T_1)}{C(T_1)} \cdot \frac{1}{T_2 - T_1},$$

   where $R(T_2)$ is the value of the resistance at temperature T_2, etc. The
   values of TCR and TCC for a specific component depend on the material
   and internal microstructure.

   Resistors of different physical construction have different thermal
   response times. This implies that two nearby resistors may be operating
   at different temperatures from each other. In addition, proximity to other

components can cause resistors to operate at different temperatures with a negative impact on circuit performance.

(a) If a 1 kΩ resistor has a TCR value of $+100$ ppm/°C, how much will its resistance change (in %) if the resistor's temperature increases by 50°C due to current flowing through nearby components? (*Note*: 1 ppm $= 1 \times 10^{-6}$).

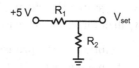

(b) In the voltage divider circuit shown, the two precision resistors have the values $R_1 = 5.00 \, \text{k}\Omega$ and $R_2 = 1.00 \, \text{k}\Omega$. What is the expected value of V_{set}?

(c) If R_1 has a TCR value of $+100$ ppm/°C and R_2 has a TCR value of $+300$ ppm/°C, how much will V_{set} change (in mV) from its nominal value if the temperature of R_2 increases by 50°C while the temperature of R_1 increases by 80°C.

(d) Suppose V_{set} is amplified by a factor of 100 and applied to a piezo actuator that expands by 40 nm for every 1 V appied. How much will the piezo's offset position change due to the thermal drift of V_{set} calculated in part (c) above?

(e) What happens if the voltage divider shown above is designed intelligently, using similar resistors for R_1, R_2 and locating both components on a circuit board so they come to the same final temperature? How much will V_{set} change (in mV) if the TCR for both R_1 and R_2 is $+100$ ppm/°C and both resistors increase in operating temperature by 50°C?

Does this problem provide some insight into why AFM electronics should be allowed to stabilize before attempting serious data acquisition?

2. If you want to achieve an elongation of 100 nm by applying 100 V to a piezo bar with a thickness of 1.5 mm, how long must it be? Assume the piezoelectric voltage coefficient is 150×10^{-12} C/N.

3. Reconsider the problem discussed in Example 7.2. Now, instead of an applied voltage, suppose the confined ceramic bar is subjected to a temperature increase. To make the problem simple, assume the temperature of the confining support remains essentially unchanged. What must be

the temperature increase to generate the same stress on the confining support? Assume the temperature coefficient α of the ceramic piezo bar is $1 \times 10^{-5}{}^{\circ}\mathrm{C}^{-1}$.

4. Estimate the capacitance between the inner ground electrode and one of the outer four electrodes of a segmented piezo tube that is 2 cm long, with a 0.5 cm outer diameter and a wall thickness of 1 mm? Assume the tube is made from a piezo ceramic material with a dielectric constant of 1200.

5. Vibration criteria VC-E states that floor vibrations should have an rms velocity less than $3\,\mu\mathrm{m/s}$ for all frequencies less than 100 Hz.

   (a) Assume that floor vibrations are primarily caused by different pieces of equipment with moving parts that undergo cyclic motion when energized. This equipment might reasonably cause the floor to oscillate sinusoidally at some frequency f (in Hz) with a constant amplitude z_o. If this is the case, what algebraic expression might be used to approximate the floor's vibration as a function of time?

   (b) If a laboratory floor is certified to meet the VC-E vibration specification, what is the corresponding amplitude of the floor vibration at 50 Hz?

6. When an AFM head is designed, it inevitably contains a mechanical loop as shown in the diagram below. Thermal stability could be an issue if the frame, the z-Stage and the xy-scanner are made from different materials. Because the Gap distance G is of order 1 nm, it is useful to write at some reference temperature $\mathrm{T_{ref}}$ that

$$G = L_o - (L_z + L_{XY}) = 0.$$

Write an expression for the gap drift $\Delta G/\Delta T$ as a function of temperature T. Assume the frame has a size set by $L_o = 10\,\mathrm{cm}$. Calculate how the gap will change for a $1{}^{\circ}\mathrm{C}$ change in temperature of the AFM head. A positive value for the thermal drift implies an increasing gap size. Use representative values found in Table 7.2.

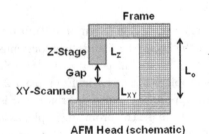

AFM Head (schematic)

A more realistic estimate must include *all* materials in the mechanical loop, including the AFM tip and the sample. AFM manufacturer's rarely quote gap drift specifications, even though this value is an important parameter required to assess the overall stability of an AFM design.

7. Experienced AFMers often can distinguish mechanical sources of noise from electrical sources just by considering the frequency range at which the offending noise appears in a signal being monitored. To make this point clear, complete the table by estimating the typical frequency range over which you might expect each noise source to appear.

Noise Source	Typical Frequency (Hz)	Mechanical/ Electrical?
Power Main		
Vacuum Forepump		
Switching Power Supply		
Wall Vibrations		
HVAC controls		
Microprocessor clock		

Chapter 8

Contact Mode AFM

In Chapter 7, we broadly discussed the essential components of an AFM system: three orthogonal positioners capable of nanometer resolution, a sharp tip mounted to a cantilever, a photodiode that measures the cantilever deflection, and a feedback loop that adjusts the position of the substrate to keep the cantilever deflection constant. With this system, it is possible to image with nanoscale precision the surface topography of a wide variety of different substrates.

In this chapter, the basic experiments that can be performed using contact mode AFM will be reviewed. A systematic framework will be provided to explain how data from various experiments can be interpreted.

Simple checks and calibrations will be suggested that are required if careful work is desired. When performing AFM experiments, it is important to remember that just because you are being careful does not guarantee that no mistakes will be made. In short, diligence in developing user-defined cross-checks at each step is required to produce convincing data.

In this chapter we discuss a number of AFM experiments performed using a static, non-oscillating cantilever. In quite general terms, this implies the tip is loaded against a sample, exerting a constant force during the course of an experiment. Such contact mode experiments should be of general interest to anyone using an AFM because (i) they constitute an important subset of experiments available to the AFM user and (ii) they lay the basic foundation for more advanced topics like dynamic AFM that rely on monitoring the cantilever motion as it oscillates in a periodic way.

8.1   Measuring Force vs. Z-displacement

A fundamental measurement with an AFM is to measure the tip-substrate force as a function of the tip–substrate separation, d. Ideally, when such a measurement is performed, useful information like the Hamaker constant, the tip–substrate force at a particular tip–substrate separation, or the indentation could be determined. In this section, the standard procedure used to analyze force vs. displacement data on hard samples is laid out in a simple way. This discussion is satisfactory for routine work. In Sec. 8.2, a more detailed approach is discussed that provides quantitative information about the tip-substrate interaction.

It would be useful if the AFM could directly characterize the tip–substrate interaction, but unfortunately it cannot. Instead, the AFM only tells you the voltage from the position-sensitive photodiode as a function of the voltage applied to the z-positioner. The challenge is to understand quantitatively the relation between the parameters you set and the parameters you measure so you can deduce physical quantities that are of interest. To this end, we discuss how the measured voltages are related to various displacements and describe the standard method implemented in AFM software that provides useful information that can be calculated even though not directly measured. Be aware there are many ways to unravel the data and many schemes have been discussed in the literature.

For the discussion that follows, it is useful to identify four lengths that play a crucial role in understanding force vs. distance data. These length parameters are identified in Fig. 8.1

(i) q, the deflection of the cantilever from some fixed position which is often taken to be the tip's equilibrium position when the substrate is far from the tip;

(ii) z, the position of the substrate's surface (i.e., the sample's surface) from an arbitrary $z = 0$ location;

(iii) z_o, the position of the undeflected tip measured from an arbitrary $z = 0$ location; and

(iv) d, the tip-substrate separation distance.

These four length parameters have units of length but are fundamentally different. The force-distance data links everything together. For example, the tip–substrate separation d (a quantity of considerable interest) is difficult to directly measure and must be inferred from changes in z coupled

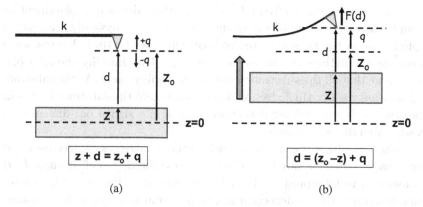

Fig. 8.1   The tip–substrate geometry. In (a), the tip is far from the substrate and experiences no force. The substrate surface is an unknown offset distance z_o from the $z = 0$ location of the z-positioner. The cantilever deflection is specified by the variable q which in this case is given by $q = 0$. In (b), the substrate has been moved closer to the tip by a coarse approach procedure and the tip is now within range of the z-positioner. Here, an assumed repulsive force $F(d)$ pushes the tip away from the substrate. In equilibrium, the cantilever deflection is $+q$ and the tip–substrate distance is d as shown.

with a simultaneous measurement of the resulting change in q. The distance z is the only parameter that is experimentally controllable since it is linearly proportional to the voltage applied to the z-positioner. Thus z is usually given by

$$z = S_z V_z, \tag{8.1}$$

where S_z (nm/V) is a scale factor often provided by the AFM manufacturer. For careful work, the precise value of S_z must be re-checked by calibration against a laboratory standard (see Chapter 9).

In Fig. 8.1, the location of $z = 0$ (with respect to the surface of the substrate) is arbitrary and unknown because it depends on many things that the user cannot control. Furthermore, z_o is quite large, 100's of microns, and its exact value depends on many uncontrollable issues like the substrate height, cantilever mounting, etc. Usually, the z-positioner has a total range of about ± 500 nm, so the surface of the substrate cannot be readily brought into close proximity of the tip. To reduce z_o to a value that is accessible by the z-positioner requires a coarse approach procedure that depends on the specific manufacturer of the AFM. For our purposes here, all we require is that a coarse approach has been successfully executed and the tip is within range of the z-positioner as shown in Fig. 8.1(b).

The parameter z is often referred to as the substrate displacement or simply displacement while the quantity d (which is physically relevant) is referred to as the tip-substrate distance. Suppose the tip is located some distance d from the substrate as shown in Fig. 8.1(a). Initially, the value of d is so large that all tip–substrate forces are negligibly small. As the substrate approaches the tip, tip deflection due to a non-zero tip–substrate force will be measured by cantilever deflection q which is a signed one-dimensional vector quantity as indicated.

After executing a coarse approach, when z increases, d decreases and eventually a tip-substrate force $F(d)$ will cause the tip to deflect (here $F(d)$ is assumed to be repulsive) through a distance q as shown in Fig. 8.1(b). In equilibrium, the tip deflection will be determined by the spring constant k of the cantilever and the tip–substrate interaction force $F(d)$. For a fixed d, in equilibrium, we have no net force acting on the tip so

$$F(d) = kq. \tag{8.2}$$

Experimentally, the AFM controls z, but as discussed in Chapters 3 and 4, the tip–substrate force is determined by d. If the interaction force is repulsive as shown in Fig. 8.1(b), evidently we must have

$$d = q + (z_o - z). \tag{8.3}$$

Since the initial position of the tip (denoted by z_o) with respect to the surface of the substrate is unknown, only the relative value of z is known as indicated by Eq. (8.3). It follows that the position where $d = 0$ cannot be defined. Many conventions for this assignment can be found in the literature and a discussion of this issue follows.

A typical force vs. z displacement experiment measures the photodiode voltage (V_{diff}, proportional to cantilever deflection) as z is systematically changed. Typical force-displacement data almost always exhibit four well-defined features as shown in Fig. 8.2.

1. The approach region is characterized by a constant difference voltage from the photodiode as z changes. If the laser beam reflected from the cantilever is carefully positioned on the photodiode (Fig. 7.3) and if a myriad of possible offset errors are small (e.g., the laser does not become misaligned during coarse approach, there is no spurious reflected light from the substrate into the photodiode, there are negligible thermal gradients, etc.) the voltage from the photodiode should be close to 0 V as indicated in Fig. 8.2 over a wide range of values for z.

2. Eventually, a rapid transition occurs when the photodiode signal abruptly changes from a constant value (in Fig. 8.2, the change is from

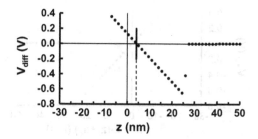

Fig. 8.2   Typical force vs. z displacement data when a tip approaches a hard substrate like SiO$_2$. The output of the position sensitive photodiode is plotted as a function of z displacement. Only one of every ten data points is plotted for clarity. The origin position for z (where $z = 0$) in this plot is arbitrary.

$0\,\mathrm{V}$ to $-0.6\,\mathrm{V}$). This designates jump-to-contact and typically only a few data points are collected in this transition region. The abrupt change in voltage indicates the tip has abruptly jumped into contact with the substrate, causing q to have a negative value. The origin of this phenomenon will be discussed in more detail in Sec. 8.2

3. As z is further changed, the tip remains in contact with the substrate and the photodiode signal increases linearly until it passes though the $0\,\mathrm{V}$ level. For the data in Fig. 8.2, this occurs when $z \approx 4.0\,\mathrm{nm}$. This condition indicates the tip has returned to its initial position and $q = 0$, indicating the cantilever deflection is zero.

4. The voltage from the photodiode finally reaches a preset value specified by the AFM user. This set-point terminates (or reverses) the scan of the z-displacement. In Fig. 8.2, the set point voltage is about $+0.36\,\mathrm{V}$.

If $q = 0$ when $z \approx 4.0\,\mathrm{nm}$, then the net force on the cantilever must be close to zero, since the $q = 0$ condition was initially set when the reflected laser beam from the cantilever was adjusted on the photodiode with the tip far from the sample. Hence it is reasonable to assume that when the output of the photodiode returns to $0\,\mathrm{V}$, the substrate is again uncompressed but the tip is now in contact. The tip–substrate separation d must then equal the interatomic spacing a_o. This realization can be used to determine z_o. It follows from Eq. (8.3) that z_o is given by

$$d = q + (z_o - z)$$
$$a_o = 0 + (z_o - 4\,\mathrm{nm}) \tag{8.4}$$
$$z_o = a_o + 4.0\,\mathrm{nm}$$

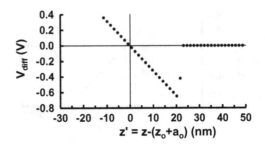

Fig. 8.3   The same data as in Fig. 8.2, except re-plotted after the z displacement is corrected by the offset z_o. The parameter $a_o \approx 0.3\,\text{nm}$ is also included to take into account the typical spacing between atoms when the tip and substrate are in contact. The origin position for z' (where $z' = 0$) in this plot is now fixed.

This allows the force vs. displacement data to be re-plotted in Fig. 8.3 so that now the tip is in contact with the substrate when $z = z_o + a_o$. This means that a new displacement variable z' can be defined such that

$$z' = z - (z_o + a_o). \tag{8.5}$$

The final modification to the data requires a conversion of the photo-diode voltage V_{diff} into the cantilever deflection q. Once this is done, the force that the tip exerts on the substrate for any value of z' can be calculated using Eq. (8.2). This final conversion is accomplished by determining a multiplicative factor S_q (nm/V) which is multiplied by the photodiode voltage to give q

$$q = S_q V_{\text{diff}}. \tag{8.6}$$

S_q often depends strongly on the laser positioning onto the cantilever, so it can be different for each AFM experiment. How is S_q determined?

For elastically hard samples, once the tip is in contact with the substrate, the cantilever will deflect as much as the substrate moves. Thus the slope of the z displacement vs. photodiode signal (assuming negligible deformation occurs after contact of the tip and substrate) allows an estimate for S_q. By analysis of Fig. 8.3, a change in z displacement of 10.0 nm produces a photodiode voltage change of 0.319 V, so $S_q = 31.3\,\text{nm/V}$. This value can then be used to convert the voltage in Fig. 8.3 into cantilever displacement as shown in Fig. 8.4.

If the spring constant of the cantilever is accurately known, say $k = 1.0\,\text{N/m}$, then the deflection of the cantilever in Fig. 8.4 can be converted directly into the force acting on the substrate using Eq. (8.2). In this

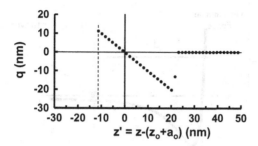

Fig. 8.4   The same data as in Fig. 8.3, except re-plotted after the photodiode voltage has been converted to q, cantilever displacement.

way we can conclude from the data in Fig. 8.3 that the maximum set-point force applied by the cantilever is $(0.36\,\text{V} \times 31.3\,\text{nm/V}) \times 1.0\,\text{N/m} = 11.3\,\text{nN}$.

Careful force vs. displacement experiments require further analysis that accounts for any possible tip/substrate deformation. This issue is discussed next in Sec. 8.2.

8.2   Force Spectroscopy

The previous section provided an overview of force vs. z-displacement experiments commonly conducted with an AFM. These experiments can be analyzed more thoroughly by taking into account the details in $F(d)$. When this analysis is carefully performed, the results are often referred to in the literature as force spectroscopy [capella99], [butt05].

AFM force spectroscopy requires an understanding of the ideas outlined in Sec. 5.7 and relies on the force diagram shown in Fig. 8.5. Initially, suppose we adjust the z-displacement to some set value as indicated by the black dot. Following Eq. (8.5), we should probably use the quantity z' when discussing displacement, but since z and z' are related by a constant offset, we simplify the notation and just use z in what follows.

The dashed line with slope k indicates the force that will develop if the tip is artificially displaced from its set z value with no substrate present (i.e., the tip-substrate forces are zero). The spring force always acts to restore the tip to its initial position and so it will have positive or negative values about the set value of z depending on the direction of q.

When the tip experiences an attractive force due to the substrate, the cantilever bends until the equilibrium condition specified by Eq. (8.2) is met. Graphically, this occurs when the dashed line with slope k intersects

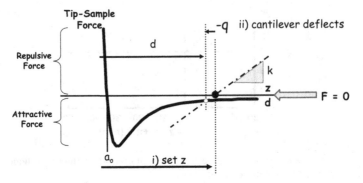

Fig. 8.5   Determining the deflection of the cantilever (q) as a function of the substrate displacement (z). For a given value of z, the cantilever will deflect by a distance q until equilibrium is reached and the forces balance. The value of the cantilever spring constant k is critical in this discussion because it determines the slope of the dashed line.

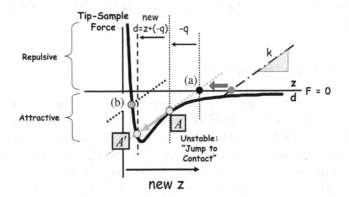

Fig. 8.6   Illustrating the jump to contact phenomenon when the tip deflection abruptly increases and the tip jumps into contact with the substrate. The region of F vs. d between A and A' becomes inaccessible.

the tip–sample force curve. At this position, the tip-substrate force at $d = z - q$ equals the restoring spring force kq. The important point to remember is that for a specified z, stable equilibrium occurs when the force vs. cantilever displacement line intersects the tip–substrate force curve.

As the substrate continues to approach the tip, the displacement of the cantilever will increase due to the ever increasing magnitude of the tip–substrate interaction force. Eventually, a situation will be reached as shown in Fig. 8.6 which illustrates the jump to contact phenomenon. At some new z value, indicated in Fig. 8.6 by the point a), the dashed line with slope k will become tangent to the tip-sample force curve at some point A. This implies that the cantilever's force vs. displacement curve no longer intersects the tip–substrate force at A, but rather at point A′.

When the tip deflects though a distance q as shown, the tip will continue to deflect until the point A′ is reached. All this occurs at the same z value. This is the jump-to-contact phenomenon and indicates instability in the tip motion due to the ever increasing tip substrate force. Once point A′ is reached, the remainder of the force vs. z displacement curve can be explored by decreasing z, as indicated schematically by the point b). The region between A and A′ is inaccessible.

The uncontrolled snap-in behavior has far reaching consequences, since on the basis of the discussions in Chapter 6, one might conclude that higher sensitivity to smaller forces can always be gained by designing cantilever probes with increasingly smaller spring constants. However, as the spring constant of the cantilever decreases, the jump-to-contact distance increases such that the probe tip will become unstable at distances many nanometers away from the substrate. In this case, the inaccessible region of the force-distance (A $\rightarrow$ A′ in Fig. 8.6) increases such that no information can be gained about the small forces of interest.

Can this jump-to-contact phenomenon be eliminated? The answer is yes, by using a stiffer cantilever. The situation is sketched in Fig. 8.7. Now, the cantilever stiffness is sufficiently high so that the force vs. z-displacement of the cantilever intersects the interaction force curve only once for any value of z. The limitation resulting from this approach is a decrease in the photodiode signal because the displacement of the cantilever is now smaller due to the increased spring constant of the cantilever.

The sequence of events illustrated in Fig. 8.6 is not what is directly observed experimentally when a Force vs. z-displacement experiment is performed. This should be evident since the horizontal axis in Fig. 8.6 is labeled by both z and d. The cautious AFM user must mentally switch back and forth between these two variables using Eq. (8.3). Experimentally, one measures the voltage out of the photodiode as a function of the substrate displacement z. What is the connection of this measurement to the more fundamental force vs. separation that dictates the cantilever's motion?

The answer to this question is given in Fig. 8.8. Figure 8.8(a) shows the analysis already discussed in conjunction with Fig. 8.6. Fig. 8.8(b) sketches what is observed experimentally.

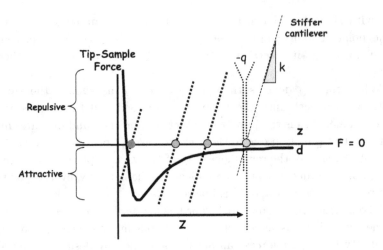

Fig. 8.7   An illustration how the jump to contact phenomenon can be eliminated if stiffer cantilevers are used. Unfortunately, stiffer cantilevers produce a smaller deflection q making accurate measurements of $F(d)$ more difficult.

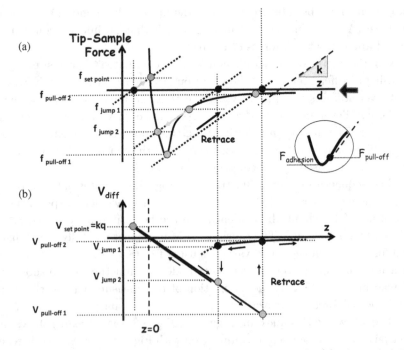

Fig. 8.8   Analysis of a Force vs. z-displacement experiment. In (a), the underlying physics behind an experiment designed to measure the tip-substrate interaction force as a function of the tip-substrate distance d. In (b), what is measured experimentally — the output voltage of the position sensitive photodiode (V_{diff}) vs. z.

Beginning at the far right of Fig. 8.8(b), if the laser spot reflected from the backside of the cantilever is carefully positioned on the photodiode, then $V_{diff} = 0$ when the tip is far from the sample. In what follows, we assume the substrate approaches the tip. Also, we assume the AFM electronics insures that the substrate will approach the tip until a set point voltage $V_{diff} = V_{set\ point}$ is recorded by the position sensitive photodiode (PSD).

The cantilever bends toward the substrate producing a voltage as shown by the black solid line in Fig. 8.8(b). In Fig. 8.8(b), this voltage is assumed to be negative, but the actual polarity will be determined by the details of the electronics controlling the AFM in use.

When the jump-to-contact condition is met, the cantilever will snap into the substrate at some z. Thus there will be an apparent discontinuity in the voltage from the photodiode given by $V_{jump\,2} - V_{jump\,1}$. This discontinuity indicates that a region of F vs. d is inaccessible due to the instability that leads to the jump to contact. As the substrate is continually displaced upward, the tip will remain in contact with the substrate. The contact mechanics discussed in Chapter 5 will govern the tip–substrate interaction and the loading of the tip against the substitute will occur.

As the substrate continues its upward movement, the tip deflection will be reversed and at some z, the cantilever position will assume its original, undeflected position ($q = 0$) observed for large tip-substrate separation. The value of z at which this happens is discussed in Eq. (8.4(a)). The tip will continually be loaded against the substrate until the condition $V_{out} = V_{set\ point}$ is met. This preset condition determines the maximum loading force in the experiment. The detailed behavior of the loading curve provides information about the elastic modulus of the tip-substrate contact.

Once the condition $V_{diff} = V_{set\ point}$ is met, the sweep in z is reversed. In some AFM controllers, a programmed time delay occurs when $V_{set\ point}$ is obtained to allow for transient behavior to stabilize. If the loading curve is performed under near equilibrium conditions, the retrace voltage from the photodiode should closely match the loading unless irreversible indentation has taken place.

The withdrawal of the tip continues until a z value is reached such that the restoring force due to the cantilever deflection is sufficient to overcome the adhesive force that develops between the tip and substrate. At this point, the tip snaps from the substrate and another discontinuity in the voltage from the photodiode ($V_{pull\,off\,1} - V_{pull\,off\,2}$) is measured as shown in Fig. 8.8(b). The substrate-tip separation will then continue to increase until a final value for z is reached. Usually, the range of the z sweep is determined by the parameters set in the z–sweep routine before beginning a Force vs. z-displacement experiment. Once the data from such an experiment (V_{diff} vs z) has been collected, the measured output is a voltage vs. z displacement and a further calibration must be performed to convert V_{diff} (in V) into a force as discussed in Sec. 8.1.

It is worth mentioning that the pull-off force is not technically identical to the adhesion force, as indicated schematically in the inset to Fig. 8.8(a). The pull-off force measured experimentally is proportional to the measured voltages

$$F_{\text{pull-off}} \propto V_{\text{pull-off2}} - V_{\text{pull-off1}}. \tag{8.7}$$

On the other hand, the adhesive force could have many contributions, as discussed in Chapter 5.

$$F_{\text{adhesion}} = F_{\text{electrostatic}} + F_{\text{vdW}} + F_{\text{capillary}} + F_{\text{chemcial}} + \cdots . \tag{8.8}$$

If it is known *a priori* that the adhesion force is dominated solely by (say) DMT contact mechanics, then we can approximate F_{adhesion} by

$$F_{\text{adhesion}}^{\text{DMT}} \simeq 2\pi R_{\text{tip}} W_{132}. \tag{8.9}$$

Approximating F_{adhesion} by the measured $F_{\text{pull-off}}$ plus knowledge of the tip radius R_{tip}, can then lead to estimates for W_{132} (see Sec. 4.2.3).

8.3   *d* vs. *z* and *q* vs. *z*

Although the connection between z, q and d is defined in Eq. (8.3), it is useful to further clarify the relationship within the context of a Force vs. z-displacement AFM experiment (often referred to as an approach curve). The issue is to clearly track five quantities listed in Table 8.1 that are all interrelated. Key points are (i) if z is varied while q is measured, you can infer d, and (ii) the measured cantilever deflection can be converted to a

Table 8.1   Important quantities in a Force vs. displacement experiment.

Quantity	Symbol
Cantilever displacement	z
Tip-Substrate gap	d
Cantilever deflection	q
Force exerted by cantilever	kq
Sample deformation	D

tip-substrate force by multiplying the cantilever spring constant k by the cantilever deflection q.

The connection between z and d often requires a model for $F(d)$. This can be illustrated by assuming the tip–substrate interaction is specified by the DMT contact model (see 5.4) which can be written as

$$F_{DMT}(d) = \begin{cases} -\dfrac{HR}{6d^2} & d > a_o \\[2ex] -\dfrac{HR}{6a_o^2} + \frac{4}{3}E^*\sqrt{R}(a_o - d)^{3/2} & d \leq a_o, \end{cases} \qquad (8.10)$$

where a_o is the relevant interatomic distance ($\sim$0.2 to $\sim$0.3 nm) and

$$E^* = \left[\frac{1 - \nu_{\text{tip}}^2}{E_{\text{tip}}} + \frac{1 - \nu_{\text{sample}}^2}{E_{\text{sample}}} \right]^{-1}. \qquad (8.11)$$

For a particular value of z, the equilibrium condition when no net force arts on the tip is specified by

$$F_{\text{DMT}}(d) = kq = k(d - z). \qquad (8.12)$$

The relationship between z and d is specified by the interaction force model, in this case Eq. (8.10). From Eq. (8.10a), for $d > a_o$ we have

$$z = d + \frac{1}{k}\frac{HR_{\text{tip}}}{6d^2} \quad d > a_o. \qquad (8.13)$$

For $d < a_o$, we have from Eq. (8.10b)

$$z = d + \frac{1}{k}\left[\frac{HR_{\text{tip}}}{6a_o^2} - \frac{4}{3}E^*\sqrt{R_{\text{tip}}(a_o - d)^{3/2}} \right], \quad d < a_o. \qquad (8.14)$$

Plots of d vs z are useful to help visualize how they are related. Equations (8.13) and (8.14) can be used to calculate d as a function of z shown in Fig. 8.9. Here we use the values $a_o = 0.3$ nm, $H = 2 \times 10^{-19}$ J, $k = 1.0$ N/m, $E^* = 5 \times 10^8$ Pa, and $R_{\text{tip}} = 5.5$ nm.

There are three distinct regions in Fig. 8.9(a) to consider. Region I is characteristic of the approach, when the cantilever is far from the substrate and $z = d$ because $q \approx 0$. Region II describes where the jump to contact and the pull off upon retraction are observed. In Region II, for a specified z, there are three possible solutions for d (marked as a, b and c). The middle solution is unstable and the tip will spontaneously jump from solution a to solution b at the slightest instability.

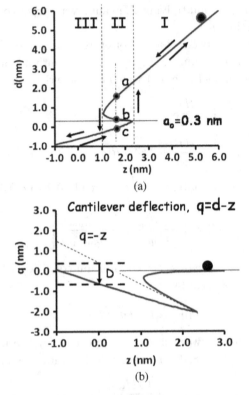

Fig. 8.9 In (a), a plot of the tip-substrate separation distance d as a function of the substrate displacement z using the DMT interaction force model specified by Eq. (8.10). The quantity a_o specifies the inter-atomic spacing when the tip contacts the substrate. The three regions of this plot I, II, and III are discussed in the text. Note that in Region II (jump to contact), the relationship between d and z is multi-valued. In (b), the cantilever deflection q is plotted as a function of z displacement. Here we use the values $a_o = 0.3$ nm, $H = 2 \times 10^{-19}$ J, $k = 1.0$ N/m, $E^* = 5 \times 10^8$ Pa, and $R_{\text{tip}} = 5.5$ nm. The dashed line illustrates q vs. z for a very hard substrate. The vertical arrow defines the deformation D as a function of z.

Finally, in Region III, the tip is in contact with the substrate. The apparent value of d is negative, indicating an indentation of the tip into the substrate (for the parameters used). Note that the position labeled $z = 0$ is arbitrary since z is controlled by the z-positioner which has an initial offset z_o relative to the tip which is usually unknown.

The results displayed in Fig. 8.9(a) can be converted into the cantilever deflection q vs. z by subtracting z from each value of d. The result is plotted in Fig. 8.9(b). The dotted line indicates the expected behavior for

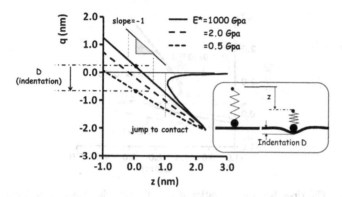

Fig. 8.10   Cantilever deflection q vs. substrate displacement z for different values of the effective elastic modulus E^* ranging between a very hard substrate ($E^* = 1000\,\text{GPa}$) to a very soft substrate ($E^* = 0.5\,\text{GPa}$). The infinitely hard substrate is usually approximated by drawing a line with slope -1 as indicated.

an infinitely hard substrate and tip, a situation in which the cantilever deflection q is equal to $-z$ since $d = 0$. The infinitely hard limit is useful in order to estimate the deformation D that is produced for a given z as indicted by the length of the vertical arrow in Fig. 8.9(b).

The ability to measure substrate elasticity is further illustrated in Fig. 8.10 where calculations similar to those in Fig. 8.9(b) are performed for different values of E^* that range between a very hard and very soft tip–substrate contact. The case $E^* = 1000\,\text{GPa}$ well approximates the case of an infinitely hard tip–substrate contact. By measuring D for a given z, the elasticity parameter E^* can be estimated. As illustrated, this measurement requires that the behavior of q (for a given z) when the tip contacts an infinitely hard substrate is known.

A final illustration of force spectroscopy that can be achieved using the force vs. z-displacement is illustrated in Fig. 8.11 which shows the sensitivity of the d vs. z data to the Hamaker constant H using the DMT model defined by Eqs. (8.10)–(8.14). For a stable system, the solid dots indicate the limiting values of z where the jump to contact occurs. Knowing the tip radius, cantilever spring constant k, and effective elastic modulus E^* of the tip–substrate, systematic measurements of the jump to contact distance could in principle be used to estimate H. The limitation of this approach is the broad range of values at which jump to contact phenomenon can occur due to instrument instabilities.

Calculations of d vs. z and q vs. z presented above are greatly facilitated by the VEDA software that will be discussed in more detail in Chapter 10.

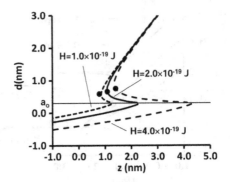

Fig. 8.11   The tip-substrate separation d vs. substrate displacement z illustrating the sensitivity to the Hamaker constant H. Here we use the values $a_o = 0.3\,\text{nm}$, $k = 1.0\,\text{N/m}$, $E^* = 2 \times 10^9\,\text{Pa}$, and $R_{\text{tip}} = 5.5\,\text{nm}$. The solid dots indicate the limiting value of z at which jump to contact can occur.

8.4   Elasticity and Adhesion Maps

The important features of the force vs. distance AFM experiment discussed in Sec. 8.2 are summarized in Fig. 8.12(a). The resulting output of the position sensitive detector (V_{diff}) is plotted in Fig. 8.12(b). Unless careful data is taken, the jump-to-contact event may not be readily apparent.

The jump to contact, substrate loading, indentation, pre-set loading force, and lift-off events should now be easily identified. The dashed sloping line indicates the indentation expected for a very hard substrate. Any deviation from this line (horizontal arrow) provides evidence for indentation and deformation.

It is useful to plot V_{diff} vs. time during the course of a force vs. displacement experiment. Such a plot can be obtained by accessing the appropriate signals from an AFM controller using a break-out interconnect box. With ready access to signals, an oscilloscope can be used to monitor the experiment. Such data might have the appearance as shown in Fig. 8.13(a) and contains a wealth of information.

If the experiment is repeated periodically at different points (i, j) on the substrate, the oscilloscope trace might resemble the plot found in Fig. 8.13(b). Once the signal in Fig. 8.13(b) is recorded, it then becomes a matter of signal processing to select the appropriate features for further analysis. If the AFM tip is positioned at pre-selected grid points (i, j) across the substrate and a force vs. distance experiment is performed at each point, a repetitive signal in the time domain with a known periodicity will result

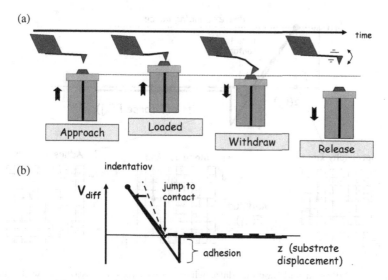

Fig. 8.12   In (a), an illustration of the classic AFM force vs. indentation experiment. In (b), the voltage from the position sensitive photodiode V_{diff} as a function of z, the substrate displacement.

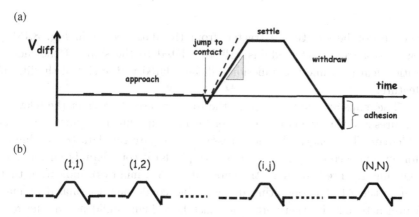

Fig. 8.13   In (a), the time dependence of the photodiode signal during a force vs. Z-displacement experiment in AFM. In (b), the result of performing force vs. displacement experiments at different points (i, j) across a substrate.

[de pablo99]. Following software analysis of this data, modulus and adhesion images of the substrate can be generated which map out the variation of elastic modulus and adhesion of the tip with nanometer-scale resolution as indicated schematically in Fig. 8.14. At each point (i, j) in the image,

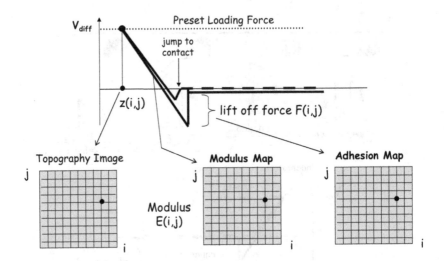

Fig. 8.14 Adhesion and modulus data, when taken across a substrate at well-defined grid points (i, j), provide position-sensitive modulus and adhesion maps of a substrate with nanometer resolution that can be correlated with a topography map of the substrate[de pablo99].

the appropriate feature extracted from the data record in Fig. 8.13(b) can be plotted. The modulus will be related to the slope of the indentation feature while the adhesion will be related to the size of the lift-off feature.

The maps are only as good as the software used to extract the relevant features and the proper execution of such an experiment requires considerable care. The length of time necessary to acquire the data ($\sim$15 minutes for an image containing 128×128 data pixels) is considerably longer than the time required to image the substrate ($\sim$3–5 minutes), indicating that unless the AFM is operating under stable conditions, thermal drift may cause artifacts. Furthermore, one must have high confidence in the reliability of the algorithm to analyze the data, since noise in the data will cause deviations from the desired waveform, resulting in maps that contain unreliable, spurious features.

An example of a careful study employing modulus maps is given in Fig. 8.15 which summarizes the results of a modulus map on a single cellulose nanocrystalline (CNC) filament deposited on mica [wagner11]. The data in Fig. 8.15 were systematically collected and contain many internal consistency checks like the calibration of the cantilever spring constant and

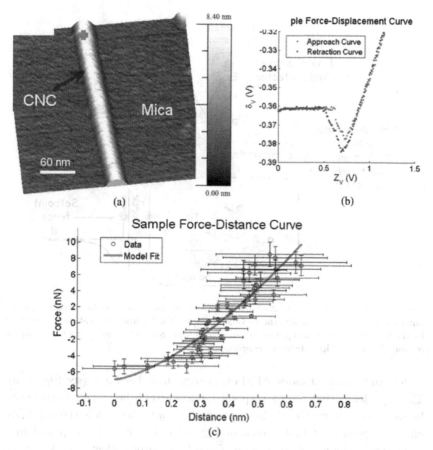

Fig. 8.15 Measuring the elastic modulus and deformation of a single cellulose nanocrystalline (CNC) filament supported on mica. (Reproduced with permission from [wagner11].)

a measurement of the tip shape before and after data acquisition. The work is notable because of the careful accounting of uncertainties during the various stages of measurement..This effort resulted in realistic error bars on the elastic modulus of the CNCs.

8.5   Contact Mode Imaging

Once it is possible to select a preset force and then systematically move the substrate toward the tip until this force is reached, the basic principles

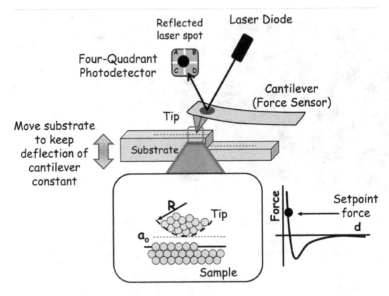

Fig. 8.16   A schematic of a contact mode AFM experiment to map the surface topography of a substrate. In contact, the distance between the tip and substrate is represented by the interatomic distance parameter a_o. AFM images are obtained using the repulsive regime of force vs. tip–substrate separation.

underlying a contact mode AFM experiment have been met (see Fig. 8.16). The key idea is that after a setpoint force is established, the substrate can be rastered beneath the tip in a systematic x-y motion while the substrate on the z positioner (not shown in Fig. 8.16) is constantly adjusted in z to maintain the specified setpoint force (cantilever deflection). As drawn in Fig. 8.16, the preset force is repulsive since the tip is in contact with the substrate. Throughout a contact mode scan, the tip–substrate separation is essentially $d = a_o$ where a_o is the characteristic atom-atom distance between the tip and substrate. In this way, the AFM software interprets the up-down motion of the z-positioner as a function of x and y to render a 3-dimensional map of the surface topography of the substrate under study.

In order to dynamically maintain a constant setpoint condition, the AFM relies on a cantilever with high optical gain to magnify small changes in the cantilever displacement q. The amplification of q by a factor of $\sim 10^2$ (see Example 7.1) is crucial for the operation of any AFM.

A number of important issues arise before performing a contact mode scan and a variety of parameters must be selected in the AFM control

Table 8.2 Typical parameters that must be set
before performing an AFM contact mode scan.

Function	Parameter
Set point force	$V_{setpoint}$
X-Y scan size	X,Y
Tip scan speed	v_{tip}
Feedback settings	K_P, K_I
Z roughness	Z_{gain}
Image size	$2^N \times 2^N$ pixels (N = 7,8,9)

program. A listing of these parameters is given in Table 8.2. Besides the
setpoint force, the scan size, the desired number of pixels in the image, the
time required to complete one line scan (the tip speed), the feedback parameters K_P and K_I, and the amplification of the voltage to the z-positioner
(Z_{gain}) must all be set in software. If the substrate is rough and has high
features, then the Z_{gain} amplification must be reduced compared to a situation where the substrate is relatively flat with low profile features. A typical
image contains $2^N \times 2^N$ pixels, where $N = 7$, 8 or 9.

As the substrate is rastered beneath the tip in a contact mode AFM
experiment, the tip can become contaminated or broken. This uncontrolled
change will influence the quality of the final image.

The proper operation of the feedback circuit shown in Fig. 7.10 is also
important. The feedback circuit continually adjusts the substrate position
in z to maintain a constant setpoint force with minimal error. The error in
the feedback will be influenced by the cleanliness of the sample, the quality
of the tip–substrate contact, and a balanced trade-off between the speed of
the tip and the response of the feedback loop.

If the substrate or tip is contaminated, the tip may stick to the substrate
at points during the x-y raster operation, producing a streaked image. If the
tip–substrate contact is not well-maintained, the image will contain artifacts
that reflect the poor contact as the scan progresses. If the speed of the tip
is too fast, the feedback loop will be unable to adjust the z-position of the
substrate quickly enough and significant errors can occur in the resulting
image.

Many techniques have been devised to verify the reliability of a contact
mode AFM image. Most AFM systems automatically acquire an error map
which plots the difference between the setpoint force and the actual force
at each pixel in the image. A quick examination of this error map allows a

visual determination of how well the constant setpoint force criterion was met while the image was acquired.

It is also customary to obtain forward and backward scans of the same region of the substrate. A significant difference between these two images provides a clear indication that the image acquired is circumspect. This comparison is greatly facilitated by calculating and then examining a "difference image" between the forward and backward scans.

Finally, two sequential images taken over the same region but at different scanning speeds should yield substantially the same results if the AFM is functioning properly.

There are distinct advantages and disadvantages to performing a contact mode scan and they should be realized. Typically, contact mode AFM imaging is only attempted on hard, clean substrates. The advantages include ease of scanning, since fewer parameters are required than for dynamic AFM imaging. Somewhat higher scan speeds can be achieved and it is possible to successfully scan rough samples with large changes in vertical topography. The disadvantages include significant lateral forces that can displace loosely bound particulates and destroy the sample and/or tip, capillary forces from adsorbed liquid layers that can cause high adhesion, and a distortion of z-height measurements when scanning softer samples due to large indentations.

There are many studies possible with contact mode AFM, and a complete recitation of all possible applications is virtually impossible. Some common measurements include substrate roughness analysis, metrology of etched features created by photolithography, thickness measurements of deposited thin films, characterization of step formation in epitaxial films, defects (e.g., pin-hole formation) in oxide growth, grain size analysis in thin films, composite material properties analysis including variations in elasticity and adhesion, and the indentation of nanoscale features at very small loads.

8.6   Non-Contact Mode Imaging

Ideally one might hope the AFM tip would behave as a non-intrusive probe, interacting with the substrate only to characterize its properties with little or no influence on the measured properties. This is often not the case. The interaction force can act to compress features of interest, resulting in erroneous estimates for their height. This is particularly troublesome when studying biological samples or soft polymers. Although the force between the tip and sample may only be 10^{-9} N, the nominal *pressure* may be $\sim 10^9$ Pa due to the small contact area. Furthermore, in contact mode AFM experiments, the tip is dragged across the substrate, causing significant lateral forces that displace loosely bound material and destroy the sample and/or tip by causing frictional wear. In careful work, these modifications to the tip require independent transmission electron microscope (TEM) studies since the scanning electron microscope (SEM) often lacks spatial resolution to resolve changes in tip structure at the nanoscale.

Some of these difficulties can be circumvented if a constant height, non-contact mode AFM scan is executed. In this mode of operation, the tip is brought close to the substrate, but does not make physical contact. The van der Waals forces bend the cantilever downward, and the setpoint force is now attractive rather than repulsive. The key distinction between contact mode and non-contact mode imaging is illustrated in Fig. 8.17.

In non-contact mode scans, the basic operation of the AFM remains the same as described in Sec. 8.5 above. As the substrate is scanned in the x and y directions beneath the tip, the z-position of the substrate is continually adjusted to maintain a constant *attractive* interaction force. The variations

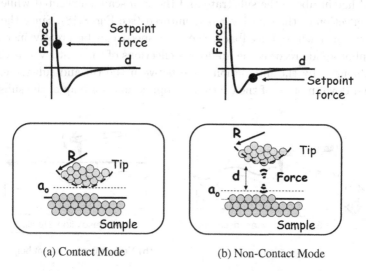

(a) Contact Mode   (b) Non-Contact Mode

Fig. 8.17 The difference between contact mode imaging and non-contact mode imaging lies in the set-point force specified before a scan is attempted. In contact mode imaging, the tip–substrate separation is essentially a_o and the cantilever deflection q is positive. In non-contact mode imaging, the tip–substrate separation is a distance $d > a_o$ and q is negative, indicating the tip is pulled toward the substrate by attractive forces.

in the z-positioner required achieving this condition as a function of x and y is then interpreted as a map of the z-topography of the substrate.

The non-contact mode of operation greatly reduces the pressure of the tip against the substrate and also reduces the lateral force that the tip exerts when scanned across the substrate. In non-contact AFM experiments, changes in the interaction force between the tip and substrate due to variations in vdWs forces can give rise to interesting image contrast since features that seem related to topography may be due to variations in the Hamaker constant across the substrate.

The practical difficulty in performing non-contact mode AFM scans is that the tip is always close to the jump-to-contact instability. Noise in the feedback circuit or unwanted vibration can cause the tip to uncontrollably snap into contact with the substrate. As a result, the non-contact image can display numerous instabilities unless the AFM system is very stable and the sample is flat. This problem is greatly reduced by vibrating the cantilever near its resonant frequency to perform non-contact dynamic AFM experiments.

To overcome these issues, non-contact mode is often performed with the feedback loop turned off. In this implementation, the tip is set to some initial height above the substrate and the x, y scan is executed while only the deflection of the cantilever is monitored (see Fig. 8.18). Since the scan is performed without feedback, the scan speed can be greatly increased. The photodiode records the up-down deflections of the cantilever as the tip is influenced by the interaction force between the tip and substrate. The up and down motion of the cantilever approximates a map of the substrate

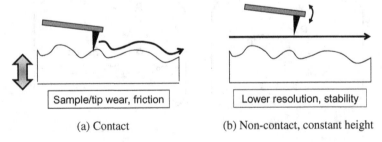

| Sample/tip wear, friction | Lower resolution, stability |

(a) Contact   (b) Non-contact, constant height

Fig. 8.18   The difference between constant force and constant height scans in AFM. In (a), the sample is moved up and down to maintain a constant cantilever deflection. In (b), the feedback is turned off and the tip is scanned at a constant height above the substrate. Variations in topography are now related to the cantilever deflection as a function of x and y.

topography. Because the magnitude of the attractive force is a non-linear function of the tip–substrate separation, the measured topography is not accurately reflected in the resulting image. This mode of operation is usually limited to flat substrates, since without feedback, any large feature on the substrate may cause a tip crash.

8.7 Lateral Force Imaging

The ability to drag a small nanometer-size tip across a substrate in a controlled way allows for the possibility of some interesting measurements of the frictional force at the nanoscale. It is well established that friction measurements at the macroscale are dominated by surface roughness as indicated in Fig. 8.19. Due to surface roughness, the true contact area might be a small fraction of the apparent contact area as illustrated in Fig. 8.19(a). Using a sharp tip obviates this problem and leads to a situation sketched in Fig. 8.19(b). Such an arrangement can potentially answer such questions as what controls friction at the nansocale? at what value of the tip–substrate separation d does friction set-in? and how does friction depend on contact area?

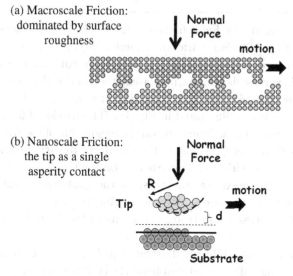

Fig. 8.19 In (a), sample roughness dominates friction measurements when two macroscale objects are rubbed against each other. In (b), a sharp AFM tip offers the possibility of reducing frictional artifacts related to sample roughness, especially if the substrate is atomically flat.

At the macroscale, the average frictional force $\langle F_{\text{friction}} \rangle$ between two objects is proportional to the normal (applied) load F_{Load}. This result is often expressed by an equation which defines the coefficient of friction μ

$$\mu = \frac{\langle F_{\text{friction}} \rangle}{F_{\text{Load}}}. \tag{8.15}$$

The coefficient of friction is a material dependent quantity that is strongly influenced by the surface condition of the two materials in close contact. Typically, $0.01 < \mu < 1$ for most materials. In principle, μ can be greater than 1, as is the case for a piece of soft rubber sliding across a flat table.

Example 8.1: A SiO_2 tip is loaded with a force of 100 nN against a GaAs substrate. A value for the coefficient of friction between SiO_2 and GaAs is found in the published literature to be 0.15. What frictional force might be expected as the tip is dragged across the substrate. According to Eq. (8.15)

$$\mu = \frac{\langle F_{\text{friction}} \rangle}{F_{\text{Load}}}.$$

With $\mu = 0.15$ and $F_{\text{Load}} = 100$ nN, a frictional force of 15 nN can be expected.

The basic geometry for a frictional force AFM-based measurement is sketched in Fig. 8.20. For a frictional force study, it is important to align the scan direction of the substrate so it is perpendicular to the long axis of the cantilever as shown. As indicated, the frictional force opposes the motion of the tip and therefore it is equivalent to an applied force acting on the apex of the tip as illustrated in Fig. 7.4. The frictional force causes the cantilever to twist. Thus, depending on the magnitude of the frictional force, one can expect the relative twist of the cantilever to vary depending on the variation of the *local* frictional force acting on the tip as it is scanned across the substrate. To maximize the twist of the cantilever, long cantilevers with long tips are preferable. By monitoring the cantilever twist as the tip is scanned across a substrate, variations in the frictional force can be mapped.

This mapping capability is illustrated in Fig. 8.21 which shows a frictional force image of oxidized graphene (OG) flakes deposited onto highly oriented pyrolytic graphite (HOPG) (Fig. 8.21(a)). The inset illustrates how the contrast in the image is proportional to the twist angle $\Delta\varphi$ of the cantilever as it is scanned across the substrate. The bright regions in

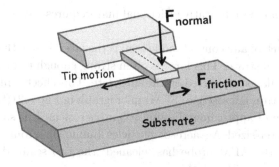

Fig. 8.20   To obtain a frictional force image, a tip must be loaded by an applied normal force as the substrate is rastered beneath it. The frictional force causes a cantilever twist that can be sensed by the lateral channel in the position sensitive photodiode.

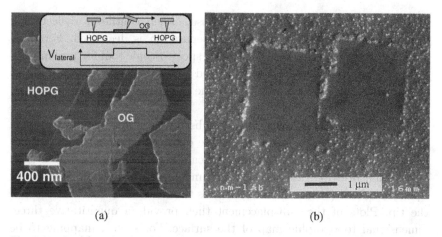

(a)   (b)

Fig. 8.21   In (a), a frictional force image of oxidized graphene (OG) flakes deposited onto HOPG. The inset shows how the signal from the lateral channel of the photodiode varies as the tip is scanned across an OG flake. (Reproduced with permission from [pandey08]). In (b), an SEM image of an InP surface covered with nanometer-sized Ag aerosol particles *after* two contact mode scans have been performed. The clearly evident rectangular patches indicate the tip has effectively swept out the Ag nanoparticles due to lateral forces. (Reproduced with permission from [junno95]).

the image indicate that the tip experiences a higher frictional force when scanned across an OG flake [pandey08].

The image plots the relative twist of the cantilever without any lateral force feedback to maintain a constant twist signal. Without the use of a feedback circuit, the image acquisition time for a frictional image can be considerably shorter than for a topographic image. To convert the frictional force image into a quantitative frictional map requires a careful calibration procedure.

A drawback of any contact mode scan is the lateral force that the tip can exert on the substrate. This force is often strong enough to dislodge weakly bound particulates supported on the substrate. This effect is illustrated in a scanning electron microscope (SEM) micrograph in Fig. 8.21(b) taken after an AFM scan of approximately $2\mu m \times 3\mu m$ across an InP surface covered with nanometer-sized Ag aerosol particles [junno95]. From the image, it is clear that the AFM probe has 'cleaned' the two scanned areas of Ag particles in a broom-like fashion.

8.8 Chapter Eight Summary

The interpretation of data was discussed when cantilever deflection q is monitored while the tip–substrate separation is decreased. Force vs. z-displacement experiments are a useful spectroscopic tool for probing tip-substrate interactions. They can be used to assess the indentation of the tip into a substrate and to estimate the adhesive force between tip and substrate.

After the tip is in contact with the substrate, contact mode topographic images of the substrate's surface can be obtained by monitoring changes in the z-positioner required to keep the cantilever deflection constant. The operational principal in contact mode imaging is to maintain a constant tip-sample force (cantilever deflection) as the substrate is rastered beneath the tip. Plots of the z-displacement then provide a quantitative three-dimensional topographic map of the surface. For such a mapping to be accurate implies the AFM feedback loop must be operational and functioning throughout the scan.

If the lateral channel of the photodiode output is monitored, a lateral force image can be obtained that maps the variation in the frictional force across a substrate. Such maps are especially useful on smooth flat substrates. Lateral force images usually do not involve a feedback circuit, so they can be obtained more rapidly than topographic images. Lateral force images nicely reveal the presence of a two-phase composition across a substrate or the presence of deposited thin flakes on an otherwise flat substrate. To convert the lateral force image into a quantitative frictional map requires further calibration procedures.

8.9   Further Reading

References to the Original Literature:

[capella99]   B. Cappella, G. Dietler, "Force-distance curves by atomic force microscopy", Surf. Sci. Rep. **34**, 1–104 (1999).

[butt05]   H.-J. Butt, B. Cappella, M. Kappl, "Force measurements with the atomic force microscope: Technique, interpretation and applications. Surface Science Reports **59**, 1–152 (2005).

[de pablo99]   P.J. de Pablo, J. Colchero, J. Gomez-Herrero, A.M. Baro, D.M. Schaefer, S. Howell, B. Walsh, and R. Reifenberger, "Adhesion Maps Using Scanning Force Microscopy Techniques", J. Adhesion, **71**, 339–56 (1999).

[junno95]   T. Junno, S. Anand, K. Deppert, L. Montelius and L. Samuelson, "Contact mode atomic force microscopy imaging of nanometer-sized particles", Appl. Phys. Lett. **66**, 3295–97 (1995).

[pandey08]   D. Pandey, R. Reifenberger, R. Piner, "Scanning probe microscopy study of exfoliated oxidized graphene sheets", Surf. Sci. **602**, 1607–13 (2008).

[wagner11]   R. Wagner, R. Moon, J. Pratt, G. Shaw and A. Raman, "Uncertainty quantification in nanomechanical measurements using the atomic force microscope", Nanotechnology **22**, 455703–11 (2011).

8.10   Problems

1. AFM images are known to produce highly accurate vertical dimensions but erroneous estimates for lateral dimensions. As a result, the width of a feature is often measured to be greater than its actual value. This result is related to the finite radius of the tip and is known as tip dilation.

   As shown in the diagram below, when a tip scans over a substrate with a feature in contact mode, the side of the tip makes contact with the feature before the tip apex. This forces the feedback circuit to prematurely respond to the feature. As a result, the feature appears to have a width greater than its actual dimension. Using the diagram below, derive an expression for the apparent width of the feature ($2b$) in terms of the tip radius (R_{tip}) and feature radius (R_{feature}). This formula will allow you to quickly estimate the feature radius if the apparent width of the feature is measured and the tip radius is known.

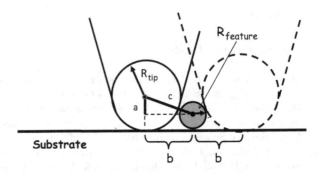

2. Assume a tip with radius 10 nm interacts with a flat substrate only through van der Waals forces. The tip is located at the end of a cantilever with a spring constant of 1.0 N/m. When the tip–substrate gap is determined to be 0.8 nm, the cantilever deflection is measured to be 0.15 nm. Estimate the value of the Hamaker constant from this data.

3. Using the parameters given in Problem 2, estimate the minimum cantilever spring constant k to avoid jump to contact. Assume $a_o = 0.3$ nm.

4. Below is a plot of typical data from an AFM force vs. displacement experiment. Use this plot to answer the questions that follow.

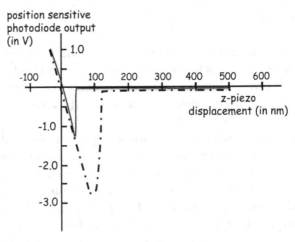

(a) From an examination of the graph, which curve (solid or dotted) represents the <u>approach</u> of the tip to the substrate?

(b) From an examination of the graph, is there any compelling evidence for long range forces acting between the tip and substrate?

(c) From an examination of the graph, estimate the jump to contact distance.

(d) It is clear the z-piezo has been assigned a specific $z = 0$ position. By inspection of the plot, state what criteria is used to determine this $z =, 0$ position?

(e) Suppose this data were taken by indenting a hard tip into a hard substrate. From an examination of this data, estimate roughly the optical sensitivity of the position sensitive detector. In other words, approximately how many nanometers of cantilever deflection produce a 1 V output from the photodiode?

(f) If a cantilever with a spring constant of 0.75 N/m was used when this data was acquired, estimate the set point force in nN.

(g) From an examination of this data, estimate the pull-off force in nN.

5. One way to estimate the Hamaker constant for a given tip and substrate is to measure the jump to contact distance of the tip to the substrate. If the spring constant of the cantilever is accurately known, the jump to contact will occur when the slope of the interaction force matches the spring constant of the cantilever. If jump to contact occurs at a tip–sample separation of 1.3 nm, use the interaction force model specified by Eq. (8.10)

$$
F_{\text{DMT}}(d) = \begin{cases} -\dfrac{HR}{6d^2} & d > a_o \\[2mm] -\dfrac{HR}{6a_o^2} + \dfrac{4}{3}E^*\sqrt{R}(a_o - d)^{3/2} & d \le a_o \end{cases}
$$

to estimate the Hamaker constant in terms of the known spring constant of the cantilever. What are the limitations of this technique for determining H?

Chapter 9

Experimental Calibrations

The experiments described in Chapter 8 may seem straightforward, but they require cross-checks and calibrations if careful work is desired. While it is often fashionable to use AFMs located in shared facilities, this practice can cause problems because other users of the equipment might adjust critical settings either in hardware or software that may negatively impact your work. Without performing cross checks, you may not suspect that the AFM data you are acquiring is uncalibrated. It is important to realize that just because you are trying to be careful doesn't mean that you will not make a mistake. In short, diligence and documentation at each step is required for your work to be accepted by others.

9.1 Calibration of a Cantilever's Flexural Spring Constant under a Normal Load

In general, quantitative AFM experiments require knowledge of the cantilever spring constant that is used to regularly relate tip deflection to applied force. While many AFM experiments are concerned with the simple flexural "up and down" motion of the cantilever, in principle there are n eigenmodes and each of the n eigenmodes has a unique spring constant k_n. In what follows, we focus on k_1, the spring constant for the first eigenmode.

In principle, a manufacturer's specifications should allow a straightforward calculation of the spring constant of a rectangular cantilever as we derived in Chapter 6. If the cantilever dimensions and the elastic modulus of the cantilever material are known, then we have

$$k_1 = k = \frac{Ewt^3}{4L^3} \tag{9.1}$$

While the cantilever width w and length L can be checked with sufficient accuracy from a scanning electron microscope (SEM) micrograph, the thickness t, which is of order $1 \, \mu$m or less, is more difficult to determine with sufficient precision to allow an accurate calculation of k from Eq. (9.1). It therefore becomes important to develop experimental techniques to calibrate k for each cantilever. The calibration is necessary because variations in the processing conditions across a Si wafer can easily cause deviations by a factor of order 2 in the spring constant between cantilevers taken from the same Si wafer.

There are four techniques that have been widely used to allow an accurate calibration of the cantilever spring constant: (i) thermal noise, (ii) Sader's dynamic method; (iii) the method of added mass; and (iv) the deflection of a reference cantilever. Each technique will be reviewed below.

9.1.1 *Thermal noise calibration*

A cantilever in equilibrium with its surroundings will fluctuate at frequencies near the natural vibrational frequencies (normal modes) of the cantilever. The thermal noise method for determining the spring constant of a cantilever requires a measurement of the displacement fluctuations of the cantilever induced by collisions with molecules in the ambient gas [hutter93][butt95]. When using the thermal method, it is important to account for contributions in the displacement due **only** to one bending mode of the cantilever. The discussion that follows assumes the first bending mode has been selected for analysis.

The thermal method of calibrating the spring constant of a cantilever relies on the thermodynamic equipartition theorem which states that in thermal equilibrium, each degree of freedom for a system will contain $\frac{1}{2} k_B T$ of thermal energy. For a cantilever, this thermal energy will reside in the various vibrational modes of oscillation. Focusing only on the cantilever's first mode of vibration, the equipartition theorem states that

$$\frac{1}{2} k_1 \langle q_1^2(t) \rangle = \frac{1}{2} k_B T \tag{9.2}$$

where k_1 is the (unknown) spring constant for the first mode of the cantilever's motion, $\langle q_1^2(t) \rangle$ is the average value of $(q_1)^2$, the average value of the square of the displacement near the resonance frequency of the first mode, measured over a very long time interval, k_B is Boltzmann's constant, and T is the temperature of the cantilever in Kelvin. A detailed analysis relies heavily on the discussion of random fluctuations introduced in Chapter 6.

The average value of the square of the cantilever's displacement, $\langle q^2 \rangle$, cannot be calculated directly from the variance of q (defined in Eq. (6.42) as the average of the **squared** differences from the mean) from a time series measurement of $q(t)$ since such a measurement will contain (i) contributions from **all** vibrational modes of the cantilever, (ii) spurious noise contributions (either vibrational or electronic), and (iii) any thermal drift that may be present.

The cantilever displacement $q(t)$ can be described by a simple harmonic oscillator with a resonance frequency ω_o driven by the force of random thermal impacts of gas atoms. The acceleration of the cantilever is given by

$$\ddot{q} = \frac{F_{\text{thermal}}}{m} - \left(\frac{\gamma}{m}\right)\dot{q} - \omega_o^2 q, \tag{9.3}$$

where m is the modal mass of the cantilever, F_{thermal} describes the driving force acting on the cantilever (in this case random impulses imparted by gas molecules), γ is the macroscopic velocity-dependent damping due to the surrounding medium (presumably air), ω_o is the angular resonance frequency of the first mode of the cantilever, and $\omega_o^2 q$ describes the linear elastic response of the cantilever.

Letting $q(t) = Q(\omega)e^{i\omega t}$ the equation of motion becomes

$$\left[-\omega^2 + i\left(\frac{\gamma}{m}\right)\omega + \omega_o^2\right]Q(\omega) = \frac{F_{\text{thermal}}(\omega)}{m}. \tag{9.4}$$

Solving for $Q(\omega)$ gives

$$|Q(\omega)|^2 = \frac{|F_{\text{thermal}}(\omega)|^2/m^2}{[(\omega_o^2 - \omega^2)^2 + \beta^2\omega^2]}, \tag{9.5}$$

where $\beta = \gamma/m$. Often Eq. (9.5) is written in terms of dimensionless frequencies as

$$|Q(\omega)|^2 = \frac{|F_{\text{thermal}}(\omega)|^2/m^2}{\left[\frac{(\omega_o^2-\omega^2)^2}{\omega_o^4} + \frac{\beta^2\omega^2}{\omega_o^2}\frac{\omega_o^2}{\omega_o^2}\right]} = \frac{f_o^4|F_{\text{thermal}}(f)|^2/m^2}{\left[(f_o^2 - f^2)^2 + (\frac{f_o^2}{Q^2})f^2\right]}, \tag{9.6}$$

where the damping factor β is equivalent to the reciprocal of the Q-factor $(1/Q)$ of the system. As discussed in Chapter 6, $Q(\omega)$ is useful to evaluate the power spectral density function PSD(ω).

Because thermal noise is random (i.e., uncorrelated over time), and since we are dealing with the first mode resonance of the cantilever, we can set $|F_{\text{thermal}}(\omega)|^2/m^2 = A_1 =$ constant, independent of frequency in Eq. (9.5). This means that any measurement of $\text{PSD}(\omega)$ can be fit to a fitting function $\text{Fit}(\omega)$ of the form

$$\text{Fit}(\omega) = \text{background} + \frac{A_1}{[(\omega_o^2 - \omega^2)^2 + \beta^2\omega^2]}, \qquad (9.7)$$

where the background term, ω_o, A_1 and β are obtained from a least squares fit to the experimental data. A typical fit to thermal fluctuation data is given in Fig. 9.1.

The quantity of interest $\langle q_1^2(t) \rangle$ in Eq. (9.2) is given by making use of Eqs. (6.49) and (6.50)

$$\langle q_1^2(t) \rangle \equiv \lim_{T \to \infty} \frac{1}{T} \int_{-T/2}^{+T/2} |q(t)|^2 \, dt$$

$$= \int_{-\infty}^{+\infty} |Q(\omega)|^2 d\omega = 2 \int_0^{\infty} \text{PSD}(\omega) \, d\omega, \qquad (9.8)$$

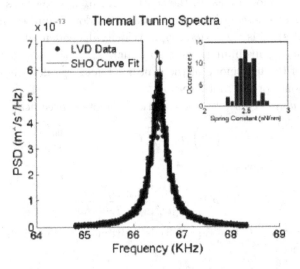

Fig. 9.1 Experimental measurement of the power spectral density of a fluctuating cantilever as a function of frequency near the first mode resonance. The solid line is a least-square fit to the theory. The inset shows a histogram for the value of the spring constant determined by repeating the PSD measurement 60 times. In this way the cantilever spring constant was determined to be 2.5 ± 0.1 N/m. Here, the velocity dq/dt rather than the displacement q was measured using a laser Doppler vibrometer. (Reproduced with permission from [wagner11]).

where T is the time interval over which $q(t)$ has been measured. The relevant integral of $\text{PSD}(\omega)$ can be calculated as follows

$$\int_0^\infty \text{PSD}(\omega)d\omega = \frac{1}{2}\int_{-\infty}^{+\infty} |Q(\omega)|^2 d\omega$$

$$= \frac{1}{2}\int_0^\infty \frac{A_1}{(\omega_o^2 - \omega^2)^2 + \beta^2\omega^2}d\omega. \tag{9.9}$$

In this last form, the integral can be evaluated using complex analysis and is equal to $\frac{\pi}{\beta\omega_o^2}$. Thus we have

$$\langle q_1^2(t)\rangle = \int_0^\infty \text{PSD}(\omega)d\omega = \frac{1}{2}A_1\frac{\pi}{\beta\omega_o^2}. \tag{9.10}$$

Using this result in the equipartition theorem, Eq. (9.2) gives

$$\frac{1}{2}k_BT = \frac{1}{2}k_1\langle q_1^2(t)\rangle = \frac{1}{2}k_1\left[\frac{1}{2}A_1\frac{\pi}{\beta\omega_o^2}\right]$$

$$\Rightarrow k_1 = \frac{2k_BT}{A_1\pi}\beta\omega_o^2. \tag{9.11}$$

Usually the frequency f rather than the angular frequency ω is the preferred experimental variable. For this reason, the second form of Eq. (9.6) is a more natural equation to use in Eq. (9.9). Correspondingly, the equation for k_1 must be modified such that

$$\omega_o \to f_o \quad A_1 \to A_1 f_o^4 \quad \beta \to \frac{f_o}{Q}$$

$$k_1 \to \frac{2k_BT}{(A_1 f_o^4)\pi}\frac{f_o}{Q}f_o^2 = \frac{2k_BT}{\pi A_1 Q f_o}. \tag{9.12}$$

Since T is known and A_1, Q and f_o are measured from a least squares fitting of the measured PSD at frequencies near the first mode of the cantilever, Eq. (9.12) allows an estimate for k_1.

Rather than measure cantilever displacement directly, usually the power spectral density of the photodiode signal $V_{\text{diff}}(t)$ is measured, which is related to the cantilever deflection by Eq. (8.6)

$$q(t) = S_q V_{\text{diff}}(t). \tag{9.13}$$

Recall that S_q can change whenever the laser is repositioned on the cantilever, so the value used for S_q must be chosen with care. The equation of motion for $q(t)$ (see Eq. (9.3)) can then be modified by including S_q so that

$$S_q\ddot{V}_{\text{diff}} + \beta S_q\dot{V}_{\text{diff}} + \omega_o^2 S_q V_{iff} = \frac{F_{\text{thermal}}}{m},$$

$$\ddot{V}_{\text{diff}} + \beta\dot{V}_{\text{diff}} + \omega_o^2 V_{\text{diff}} = \frac{F_{\text{thermal}}}{(S_q m)}. \tag{9.14}$$

The equation for $V_{\text{diff}}(\omega)$ becomes

$$|V_{\text{diff}}(\omega)|^2 = \frac{|F(\omega)|^2/(S_q^2 m^2)}{[(\omega_o^2 - \omega^2)^2 + \beta^2\omega^2]}. \tag{9.15}$$

It follows that $A_1 \to A_1/S_q^2$ in Eq. (9.12), so that if the power spectral density of the photodiode voltage is measured, the corresponding expression for k_1 will read

$$k_1 = \frac{2k_B T}{\pi A_1 Q f_o} S_q^2. \tag{9.16}$$

The result for k_1 turns out to overestimate of the actual spring constant k for the first mode since the theory (Eq. (9.9)) treats the cantilever as a point mass suspended from an ideal spring. This is only a convenient approximation and in careful work the result for k_1 must be modified to include the cantilever's actual mode shape [butt95]. These considerations require an accounting of the difference between the z-deflection (what is calculated) and the change in slope (what is measured) at the end of the cantilever. An additional modal correction factor is needed to convert k_1 into the actual spring constant k

$$k = \frac{k_B T}{\langle q^2 \rangle} = \frac{12}{\alpha_1^4} \frac{k_B T}{\langle q_1^2 \rangle} = 0.971\, k_1, \tag{9.17}$$

where $\alpha_1 = 1.8751$ [cook06].

9.1.2   *Hydrodynamic damping calibration*

J.E. Sader developed a method of calibrating the spring constant of a cantilever that relies on a theory that predicts the shape of the response as the driving frequency of a cantilever is swept through its resonance [sader99], [sader98]. The method discussed below applies to the spring constant of a rectangular AFM cantilever and requires knowledge of the appropriate hydrodynamic damping of the cantilever, either for operation in air or in a fluid. From a measurement of the cantilever's response, the spring constant k_1 of the first mode can be determined using the equation

$$k_1 = \frac{m}{m_c} \frac{\pi}{4} \rho w^2 L Q \omega_o^2 \Gamma_i(\omega_o), \tag{9.18}$$

where m/m_c is the ratio of the modal mass of the cantilever to m_c, the actual mass of the cantilever, L and w are the length and width of the rectangular cantilever, respectively, ρ is the density of the fluid (usually air) surrounding the cantilever, ω_o and Q are the (angular) resonant frequency and quality factor of the fundamental resonance peak, respectively, and Γ_i is the imaginary part of the hydrodynamic damping function which depends on the Reynolds number for flow in the surrounding fluid. An evaluation of Γ_i requires knowledge of the cantilever dimensions, the density and the viscosity of the fluid surrounding the cantilever. Note that for a rectangular cantilever, the factor $m\pi/(4m_c)$ equals the number 0.1906 since, $m/m_c = 0.2427$.

The quantities L and w are usually known from manufacturer's specifications and can be cross-checked using SEM micrographs. The parameters ω_o and Q are determined by fitting the lineshape of the cantilever response as the driving frequency ω is swept through ω_o. In this case, F_{thermal} in Eqs. (9.3)–(9.6) must be replaced by $F_{\text{drive}} = F_o \sin(\omega t)$. Alternatively, values for ω_o and Q can also be derived from measurements of the thermal power spectral density of the cantilever. Evaluation of the hydrodynamic damping Γ_i is facilitated by an easy to use web-based application [hydro-damping].

The density ρ of the fluid surrounding the cantilever appears in Eq. (9.18). For the case of a gas, the density depends on pressure and so the pressure must be taken into account in careful work.

As an example, consider the case of air. The density of air can be estimated assuming the ideal gas law. For n moles of gas at a pressure P, with a volume V at a temperature T

$$PV = nRT \tag{9.19}$$

where R is the ideal gas constant equal to $8.314 \text{ J mol}^{-1} \text{ K}^{-1}$. The density ρ of a gas is given by

$$\rho = \frac{n}{V} M \tag{9.20}$$

where M (kg mol^{-1}) is the molar mass of the gas surrounding the cantilever. Combining Eqs. (9.19) and (9.20) leads to an equation for ρ (kg/m^3) which

Table 9.1　Densities of a few common gasses at $20°$C and standard atmospheric pressure.

Gas	Molar Mass M (kg mol^{-1})	Density at STP (kg/m^3)
Air	0.029	1.20
Helium	0.004	0.17
Nitrogen	0.028	1.16
Argon	0.040	1.66

varies pressure linearly with P

$$\rho = \frac{P}{RT}M. \tag{9.21}$$

Since most AFM measurements are performed in air, it is useful to evaluate Eq. (9.21) for this particular case. The molar mass of air is typically taken as 0.029 kg mol^{-1}. Taking standard atmospheric pressure at sea level as $P = 1.01 \times 10^5$ Pa and room temperature as $T = 20°$C (293 K), $\rho_{air} = 1.20$ kg m^{-3}.

For reference, Table 9.1 gives values for the density of a few other common gases that are sometimes used to maintain an inert atmosphere around an AFM.

Clearly careful AFM measurements made at elevations well above sea level will require an independent measurement of P when evaluating Eq. (9.21). Also, calibrations performed in purified gases like argon or helium requires the use of the proper molar mass M as well.

9.1.3　Calibration using a known spring constant

The calibration of a cantilever with an unknown spring constant can be accomplished by using an array of reference cantilevers of known spring constant [gates07]. The basic idea is sketched in Fig. 9.2 which shows a reference spring with known spring constant k_{ref} that is compressed by a reference mass m through a measured distance Δz_1. If the reference spring is then attached to a spring with **unknown** spring constant k_c, the compound spring system has an effective spring constant k_{eff}. If the same mass now compresses the two spring system by a measured distance Δz_2, it follows that

$$mg = k_{eff}\Delta z_2 = k_{ref}\Delta z_1. \tag{9.22}$$

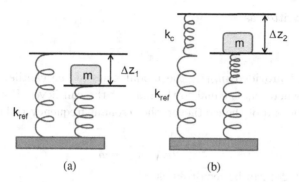

Fig. 9.2 Measuring the deflection of a reference spring with known spring constant k_{ref} provides a way to calibrate the unknown spring constant k_c of a cantilever. The method of calibration relies on an accurate measurement of the two deflections Δz_1 and Δz_2.

Because the compound system is formed by two springs in series, it has an effective spring constant k_{eff} given by

$$\frac{1}{k_{eff}} = \frac{1}{k_c} + \frac{1}{k_{ref}}. \tag{9.23}$$

It is then easy to derive an expression for the unknown spring constant in terms of the reference spring and the measured displacements

$$k_c = \frac{k_{ref}}{\left(\frac{\Delta z_2}{\Delta z_1} - 1\right)}. \tag{9.24}$$

Replacing the springs by cantilevers adds a bit of complexity since the precise point of contact at which the unknown cantilever presses against the calibrated cantilever must be determined, but the basic idea of measuring the deflection of an unknown cantilever when pressed against a cantilever with a known spring constant remains [gates07]. Repeating the measurements for a few reference cantilevers also increases the reliability of the final result.

9.1.4 *Added mass calibration*

Lastly, it is also possible to attach a known mass to a cantilever and measure the resonant frequency shift [cleveland95]. As discussed in Chapter 6, a point mass model can be used to discuss many features of the cantilever motion provided an effective cantilever mass is substituted in relevant formulas. So for example, the simple formula for the resonant frequency of a cantilever written as

$$f_o = \frac{1}{2\pi}\sqrt{\frac{k}{m_{\text{eff}}}} \tag{9.25}$$

can be used provided m_{eff} is the modal mass of the cantilever equal to $0.2427\, m_c$ where m_c is equal to the mass of the cantilever. If a mass m is added to the end of the cantilever, the resonant frequency will shift to

$$f = \frac{1}{2\pi}\sqrt{\frac{k}{m_{\text{eff}} + m}}. \tag{9.26}$$

Equation (9.26) can be rewritten as

$$m = \frac{k}{4\pi^2}\frac{1}{f^2} + m_{\text{eff}} \tag{9.27}$$

and a plot of the mass added (m) vs. $1/f^2$ should yield a straight line with a slope given by

$$\text{slope} = \frac{k}{4\pi^2}. \tag{9.28}$$

A least squares fit of data to Eq. (9.27) will yield an estimate for the slope from which k can be calculated.

The added-mass technique requires a number of additional processing steps. The technique is implemented by touching the end of a cantilever to a bed of micron-size powder, usually after experimental AFM work has been completed. Tungsten powder is used to advantage in this application because of its high mass density (19.25 gm/cm^3). A measurement of the diameter of the particle(s) adhering to the cantilever is also required. In some instances, a thin metal film is evaporated onto the end of the cantilever in place of the attached particles. If a calibrated thickness monitor is used during the evaporation, it is then possible to accurately estimate the added mass.

9.1.5 *Cantilever calibration benchmarks*

A survey of the literature indicates the two most used calibration techniques for cantilever calibration seem to be based on the equipartition theorem and hydrodynamic damping. In commercial AFMs, one or both of these techniques is often automatically executed to provide reliable estimates for k. Recently a study was reported that directly compared the spring constants determined by these two methods [cook06]. The results are given in Fig. 9.3 and show that both techniques give consistent estimates for the

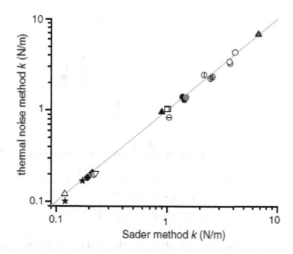

Fig. 9.3 A direct comparison between the thermal fluctuation calibration of cantilever spring constants and the hydrodynamic method, for cantilevers spanning two orders of magnitude in k. If the two calibration methods agree, all the data should lie on the solid line. (Reproduced with permission from [cook06]).

spring constants of rectangular cantilevers spanning about two orders of magnitude in k.

Another study has been reported in which spring constants of the same set of rectangular and soft, V-shaped cantilevers have been independently calibrated in eight different AFM labs using ten different AFMs to assess the reproducibility of cantilever calibration from lab to lab [tereit11]. By comparing the calibration from a single AFM to the mean of the results from ten AFMs, Sader's method was found to be more accurate. Representative results from this study are provided in Fig. 9.4.

In Fig. 9.4, the individual cantilevers are labeled MLCT1 through MLCT5 and MSCT1 through MSCT5. While each lab agrees on the general trend in the cantilever spring constant, the spread in the calibration of any one cantilever indicates about a 10% variation between the values of k determined in different labs, indicating that relative errors between different AFMs are present.

9.2 The Photodiode Calibration Constant

If the free end of the cantilever is displaced by an amount q (in nm), it becomes useful to know the corresponding V_{diff} that the photodiode

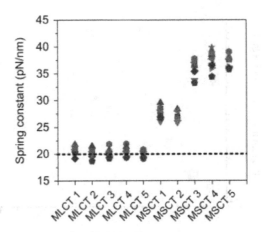

Fig. 9.4   The results of an accuracy check on the calibration of the same ten rectangular cantilevers calibrated in different labs using different AFMs. This data set was obtained using the Sader method of calibration. (Reproduced with permission from [te reit11]).

generates (see Fig. 7.3). The answer will depend on the exact position of the laser beam on the cantilever, and hence a calibration must be repeated every time the laser spot is readjusted. Most commercial AFMs have integrated routines in which this calibration is automatically performed, but it is useful to outline the various steps involved as illustrated in Fig. 9.5. The basic experiment is shown in Fig. 8.3.

In what follows, we refer to the co-ordinate system established in Fig. 9.5(a). In Chapters 2 and 3, the separation between the tip apex and the substrate was specified by d, and we maintain this identification here. Let the initial distance between a very hard substrate and the cantilever tip with no interaction forces equal z_o. Whenever the tip is deflected upwards by an amount q, we assume the deflection is positive $+q$. If the tip is deflected downwards by an amount q, we assume the deflection is negative $-q$. These definitions are the same used in the discussion of Eq. (8.3) and they allow the tip–substrate separation distance d to be defined by the motion of the substrate's surface (specified by z)

$$z + d = z_o + q, \tag{9.29}$$

where z_o is an unknown offset distance that specifies the location of the undeflected tip from the arbitrary $z = 0$ of the z-positioner (see Fig. 8.1). If the tip is far from the substrate, $q = 0$ and we have $d = z_o - z$ as in Fig. 9.5(a).

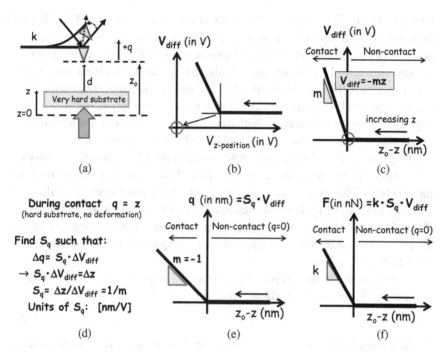

Fig. 9.5 Determining the photodiode calibration constant S_q is a multi-step process. The goal is to convert photodiode voltage (V_{diff}) vs. z-positioner voltage ($V_{z\text{-position}}$) into force vs. displacement. Most commercial AFMs now have an automatic routine which determines S_q. In this figure, the symbol m is used to designate a slope, not a mass.

By actuating the z-positioner, z is made to increase as the substrate moves closer to the tip. This simply means $(z_o - z) \to 0$ with increasing z. Here, z can range from 10's to 100's of nanometers. A tip–substrate interaction will cause the tip to deflect. Initially, the cantilever should bend toward the substrate ($-q$) by a small amount due to the attractive van der Waals forces. When the tip contacts the substrate, a repulsive force develops and the cantilever deflection is reversed. As z increases, the cantilever is restored to the $q = 0$ position and then bends upwards, with q becoming positive. Meanwhile, the laser beam reflected from the cantilever moves across the photodiode due to the cantilever deflection, causing the voltage from the photodiode (V_{diff}) to change in a systematic way. This process was previously discussed in Fig. 8.2.

The characteristic shape of the data acquired by the AFM controller is sketched in Fig. 9.5(b) which shows how V_{diff} varies as a function of the voltage applied to the z-positioner. The data at this point is very specific to each AFM. The shape of the data curve can be displaced or flipped or inverted from that shown, since the voltage to the z-positioner and the polarity of the photodiode voltage depends on the AFM electronics. For this reason, such data is generally not suitable for publication. Furthermore, the speed at which the data is acquired is an issue. If the substrate approaches the tip too quickly, cantilever instabilities may result upon contact. If the approach is too slow, thermal drift may cause a problem and a withdrawal of the tip may not mirror the data obtained during approach.

To produce meaningful data, the voltage the AFM controller sends to move the z-positioner (the x-axis in Fig. 9.5(a)) must be converted to nanometers. This conversion requires that the z-positioner is calibrated according to Eq. (8.1). It is also customary to shift the data so that the condition $(z_o - z) = 0$ corresponds to the tip-substrate contact event described by $q = 0$. This shift allows the raw data to be plotted as shown schematically in Fig. 9.5(c).

The slope m (units V/nm) of the contact region in Fig. 9.5(c) can now be measured and the photodiode voltage can be written as

$$V_{\text{diff}} = m(z_o - z) \tag{9.30}$$

Any meaningful force vs. displacement data requires a measurement of the cantilever deflection in meters rather than the measured photodiode signal in volts, so a plot similar to that in Fig. 9.5(c) is not particularly useful. To convert V_{diff} to q is easy if the substrate and tip are both very hard. In this case, the deformation of the tip-substrate contact is assumed negligibly small. This implies q must then equal z, the distance the z-positioner has moved.

The conversion of V_{diff} to q is equivalent to identifying a multiplicative factor S_q such that

$$\Delta q = S_q \Delta V_{\text{diff}} = \Delta z. \tag{9.31}$$

When V_{diff} is multiplied by S_q, the data will take the form shown in Fig. 9.5(e), where now the contact region of a q vs. z plot will have a new slope m' that must equal -1. From Eq. (9.31), this implies that the proper choice for the sensitivity S_q (in nm/V) is given by

$$S_q = \frac{\Delta z}{\Delta V_{\text{diff}}} = \frac{1}{m} \tag{9.32}$$

where m is just the slope that was measured in Fig. 9.5(c).

Once V_{diff} is converted to nanometers, it can be further converted into a force by multiplying with the known spring constant k of the cantilever. Recall that k is determined using one of the procedures already discussed in Sec. 9.1 above. This final conversion step is indicated in Fig. 9.3(f) where now the slope of the force vs. displacement data is equal to k. The data plotted in this form is now suitable for publication since it can be directly compared to similar data obtained with other AFMs and with different cantilevers.

This calibration process converts V_{diff} to a tip–substrate force which is a quantity of considerable interest. If the hard substrate is removed without disturbing the alignment of the laser on the cantilever, then any substrate of interest can be inserted for further study with the assurance that the photodiode signal can be converted to a force.

9.3 Calibration and Orthogonality of the X and Y Scanners

It is important to regularly check the X and Y scanners for calibration, linearity and orthogonality. While commercial AFMs provide an initial calibration prior to shipment from the factory, these values may change over time, either through changes in the X and Y scanners due to use (or abuse) or through deliberate or inadvertent adjustments by other users, a common occurrence in multi-user facilities. Such events result in wasted time for all until the change in calibration is detected. A strategy to avoid these problems is to devise easy-to-implement calibration procedures that can be quickly performed before serious work begins.

The best strategy to achieve an X and Y calibration is to have available a square calibration grid with periodic features as illustrated in Figs. 9.6(a)

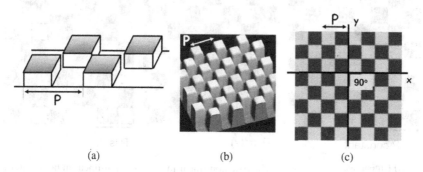

(a)   (b)   (c)

Fig. 9.6 A checkerboard substrate useful for checking the calibration of the X and Y scanners. The periodicity P of the repeating pattern is chosen to match the range of X and Y that will be calibrated. The orthogonality of the X and Y scan directions can also be determined by measuring the relative orientation of the features as shown in (c).

and (b). The calibration grid must be stored in a clean environment (like a desiccator) and cannot be left lying on a table top, for it will surely become covered (and hence useless) with dust and particulates from the air.

The periodicity of the structure is characterized by a parameter P which measures the repeat distance as indicated (see Fig. 9.6(a)). In principle, the repeat distance of the pattern in the substrate is known and is as accurate as the photomask used to create it. For calibration purposes, it is useful to have a number of such calibration grids which span a range of periodicities so the calibration of the X and Y scanners can be accurately checked at different length scales.

Figure 9.6(c) is a top-view AFM image of the periodic structure useful for calibration purposes. From such an image, two quantities can be checked. First, the periodicity can be measured and compared to the known periodicity of the pattern etched into the substrate. In this way, the calibration of the X and Y scanners can be checked. Secondly, the periodicity can be measured as a function of position. If the X and Y scanners are linear, the periodicity should be independent of position. If an X–Y scanner is non-linear, then images similar to those shown in Fig. 9.7 will be observed where the size of the periodic feature varies with position. Knowing how the periodicity changes as a function of x and y is a first step toward devising correction functions to linearize the scan. Once the functional form of the non-linearity is known, a correction in the scanning software can be made by applying a non-linear voltage to the scanners to produce a linear scanner

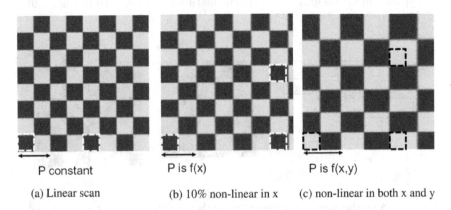

P constant	P is f(x)	P is f(x,y)
(a) Linear scan	(b) 10% non-linear in x	(c) non-linear in both x and y

Fig. 9.7   Simulated AFM scans to illustrate (a) a linear X-Y scanner; (b) a scanner that is non-linear in x, and (c) a scanner that is non-linear in x and y. Calibration requires a measurement of the dimensions of the repeating feature throughout the image.

movement. This correction often requires vigilant re-calibration since the precise non-linear voltage needed to correct for non-linear motion changes as the X and Y scanners age.

Once X and Y are calibrated, a third important check to perform is for orthogonality in the X and Y scanners. Orthogonality is different from linearity, since the latter assumes displacement is linearly proportional to applied voltage while the former requires that a displacement in X produces no unwanted displacement in Y. For piezo-tube scanners with four outer electrodes (Fig. 7.7), the ability to independently scan in both X and Y cannot be assumed since cross-coupling of motion in X and Y is a definite possibility. As shown in Fig. 9.6(c), it is possible by a simple measurement to check whether X and Y scan directions are at right angles by measuring the angle between two fiducial lines similar to those that are dawn on the image.

9.4 Calibration and Non-Linearity of the Z Positioner

The calibration of the Z positioner is perhaps the most critical calibration for an AFM since most AFM studies ultimately are performed to obtain estimates of z heights. The best strategy to achieve this goal is to purchase a trench calibration pattern with periodic trench-like features of known height as illustrated in Figs. 9.8(a) and (b). Usually, trenches of height h are etched into a SiO_2 substrate using photolithographic techniques. It is important to select a step height to match a particular application and for this reason, a substrate containing trenches of known heights (say $h = 25$ nm; 100 nm; 500 nm; 1000 nm; 1500 nm) are commercially available. For example, if you are measuring nanoparticles supported on a flat substrate with diameters of 10 nm, the use of a calibration pattern with trench heights of 1500 nm would of limited use to calibrate your nanoparticle height measurements.

Substrates with sub-nanometer steps of known height are also of interest. Often, these calibration substrates are produced by annealing single crystal substrates with nominal miscuts of a few degrees from a preferred crystallographic direction. After annealing, such miscut substrates tend to show a regular and ordered stepped morphology with clearly defined terraces that can be readily observed in AFM. The height difference between terraces can be related with high precision back to the known dimensions of the unit cell.

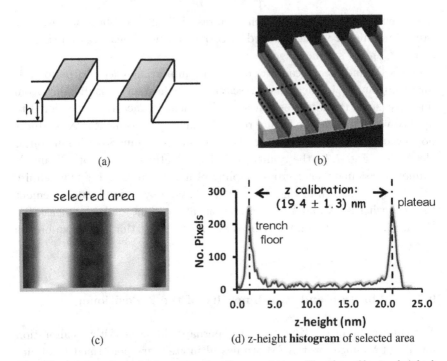

(a)

(b)

selected area

(c)

(d) z-height **histogram** of selected area

Fig. 9.8   A substrate useful for calibrating the z-positioner. Trenches of known height h are etched into a SiO$_2$ substrate as shown in (a) and (b). In (c), a selected region of the substrate in (b) is selected for further analysis. By forming a histogram of heights in the selected region, a histogram of the number of pixels with a specified z-height as shown in (d) can be obtained. This plot allows an accurate assessment of the apparent height of the trench-like features. The widths of the peaks in the histogram provide information about the accuracy of the calibration.

By performing an AFM image of a Z calibration pattern over a selected region indicated by the dotted square in Fig. 9.8(b), an AFM topograph (Fig. 9.8(c)) can be obtained. Once an AFM image is obtained, most commercial scanning probe software has the ability to (i) subtract a global plane from the image to provide a constant reference plane and (ii) plot a histogram of pixel heights as shown in Fig. 9.8(d).

For the image shown, the histogram should have two distinct peaks and the distance between these two peaks can be accurately measured. Furthermore, as evident in Fig. 9.8(d), each peak will have a statistical width which reflects the inherent surface roughness of the plateaus and the bottom of the trenches. The roughness restricts the accuracy of any height measurement and the width of the histogram peaks can be used to estimate the uncertainty for the difference in z-heights using standard methods for combining errors [baird95], [squires01]. This approach to z-calibration is much preferred over the quicker and less accurate approach of locating a single line on the image that crosses the trenches and then estimating the height difference from the profile features measured along the line.

It is often useful to check the linearity of the z-positioner over a range of z-values. A special calibration sample is required. Often, an atomically flat substrate (like HOPG) with flat terraces offset by a known step height (0.34 nm) can be employed as shown in Fig. 9.9(a). The difficulty with this approach is that a region of the substrate with a sufficient number of steps to be useful for calibration often requires a tedious search. Alternatively, a grid-like substrate containing features with a known height offset Δh, similar to that sketched in Fig. 9.9(b) could be used. The fabrication of such a grid is technically challenging. A more systematic approach is suggested in Fig. 9.9(c) where a calibration pattern similar to that shown in Fig. 9.6 is mounted on a wedge of known angle α [eaton01]. Knowing α allows a

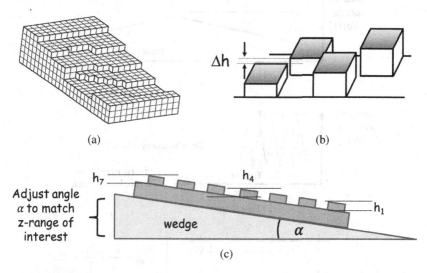

(a)   (b)

(c)

Fig. 9.9   Checking the non-linearity of the z-positioner requires a special substrate with features of known height offset from each other by a specified distance. In (a), a sketch of atomic step heights in a substrate. In (b), a grid-like substrate containing features with a known height offset Δh. In (c), a checkerboard grid substrate with uniform height features, similar to that shown in Fig. 9.6. By supporting the substrate on a wedge cut to a known angle α, an array of features off-set from each other by a known height can be obtained.

calculation of the relative separation in z of the different features which are all of the same height when $\alpha = 0$.

By performing an AFM image of this tilted substrate, the various z-heights of the features $(h_1, h_2, \ldots, h_n)$ can be determined. If the z-positioner is non-linear with z, a plot of these measured step heights will reveal a systematic dependence on z. By devising a least-squares analytical fit to the observed variation, the voltage to the z-positioner can be corrected by an analytical function to produce linear scans in z.

9.5   Scanning Issues

It is useful to discuss the time variation of the scanning voltages sent to the X and Y scanners in order to execute an AFM scan. As will become clear, the time required to execute one scan line in an AFM image has a direct impact on the feedback settings of the AFM controller. Figure 9.10

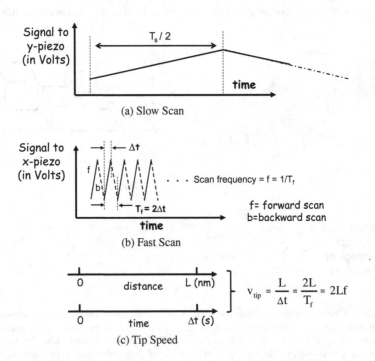

(a) Slow Scan

(b) Fast Scan

(c) Tip Speed

Fig. 9.10   The time dependence of various signals sent to the X and Y scanners in an AFM. In (a) the variation of the voltage to the slow scan (Y-scanner) as a function of time. In (b) the variation of the fast scan (X-scanner) as a function of time. In (c), calculating the tip speed while executing an AFM image.

lays out the important parameters that must be chosen when executing an AFM image. In what follows, it is useful to distinguish between the slow scan direction (usually taken as the y direction) and the fast scan direction (usually taken as the x direction).

AFM controllers often use a computer with a digital to analog board to output a continuous voltage to the x and y positioners to control their motion. The AFM software typically will accept inputs from the user to control the scanning conditions and handle image acquisition. For convenience, the time dependence of the voltage in Fig. 9.10 is represented by a smooth curve. Typically, the voltage outputs to the X and Y scanners are generated digitally, so that small steps may be observed if the scanning voltages are accurately measured.

The important parameters are the times T_s and T_f that control the slow and fast scans, since the choice of these parameters impacts AFM scanning by determining the time required for image acquisition.

The choice of T_f in Fig. 9.10(b) is most relevant. Often the scanning software will request an input (in Hz) for the scan speed. This input then governs the scanning signal sent to the X-scanner. As indicated in Fig. 9.10(b), assume the AFM is scanning in the forward direction during a time interval Δt, i.e., one forward scan requires an amount of time Δt to complete. Unless there is a pause executed at the end of the forward line scan, an equivalent time Δt will be required to execute a backward scan. The period of the output voltage to the x-positioner is then $T_f = 2\Delta t$ and the X-scan (fast) frequency can then be defined as

$$f_f = \frac{1}{2\Delta t} = \frac{1}{T_f}. \tag{9.33}$$

If $\Delta t = 2$ s, then from Eq. (9.33), $f_f = 0.25\,\text{Hz}$.

The slow scan frequency is usually determined by requiring that the final image contain $2^N \times 2^N$ pixels, where typically $N = 7$, 8 or 9. Thus if the software acquires 2^N data points per line in a time Δt, then to have a $2^N \times 2^N$ pixel image we must require that

$$\frac{T_s}{2} = 2^N \Delta t = \frac{2^{N-1}}{f_f}. \tag{9.34}$$

Equation (9.34) essentially determines the slow scan time T_s in terms of the fast scan time Δt. The total time for a forward and backward image will then just be T_s. From Eq. (9.34) we then have

$$T_s = 2^{N+1} \Delta t = \frac{2^N}{f_f}. \tag{9.35}$$

Table 9.2   Estimates in seconds for the minimum time to acquire forward and backward AFM images, each containing $2^N \times 2^N$ pixels.

f_f, fast scanning frequency	Δt	T_{total}; $N = 7$ (128 × 128 pixels)	T_{total}; $N = 8$ (256 × 256 pixels)	T_{total}; $N = 9$ (512 × 512 pixels)
10 Hz	0.05 s	12.8 s	25.6 s	51.2
5 Hz	0.10 s	25.6 s	51.2 s	102.4 s
1 Hz	0.5 s	128 s	256 s	512 s
0.5 Hz	1.0 s	256 s	512 s	1024 s

Thus an estimate for the total image acquisition time $T_{total} = T_s$ to acquire a forward and backward image can be obtained from Eq. (9.35).

Table 9.2 is a quick look-up table that gives some typical image acquisition times based on these estimates. These are minimal times since a specific AFM system may require additional delays introduced by the control algorithms.

These times can also be used to estimate the scanning velocity of the tip. If the image being acquired is square with dimensions $L \times L$, then the average velocity of the tip across the substrate v_{tip} will be given by

$$v_{tip} = \frac{L}{\Delta t} = 2 f_f L. \tag{9.36}$$

Thus if the scanning frequency is 5 Hz and the image size is 100 nm × 100 nm, then the tip velocity must be near 1000 nm/s. Knowledge of the tip velocity becomes important when simulations of AFM experiments are performed as discussed in the next chapter.

The *orientation* of the X and Y scanning directions with respect to the substrate is a controllable parameter that can be adjusted. If a new coordinate system $(x\prime, y\prime)$ is defined to be rotated by an angle θ from the original coordinate system (x, y), then scanning along $x\prime$ and $y\prime$ can be performed by defining a rotation matrix R that accounts for the rotation through the specified angle θ. Thus if it is desired to rotate the scanning directions thru an angle θ, the software must simply output new voltages $V\prime_x$ and $V\prime_y$ to the x and y positioners that are related to the original voltages V_x and V_y by

$$\begin{pmatrix} V'_x(t) \\ V'_y(t) \end{pmatrix} = R \begin{pmatrix} V_x(t) \\ V_y(t) \end{pmatrix} = \begin{pmatrix} \cos\theta & -\sin\theta \\ \sin\theta & \cos\theta \end{pmatrix} \begin{pmatrix} V_x(t) \\ V_y(t) \end{pmatrix}. \tag{9.37}$$

The rotation of scanning direction is used in two ways: (i) to align the fast scan axis along a specific direction in the substrate to minimize the

z-height variation due to a tilt angle between the tip and substrate, or (ii) to guarantee that step edges are traversed in a perpendicular rather than a parallel direction. It is worth commenting that even though most discussions of AFM tend to depict the angle between the tip axis and the substrate as 90°, this is hardly ever the case since inevitably the tip will always be oriented at some angle with respect to the substrate. Unfortunately, this tilt cannot be easily removed by a fine adjustment of either the tip or substrate and it will appear in any 3-dimensional rendering of an AFM image, often causing small features to be obscured. Most AFM image-rendering software eliminates the substrate tilt (sketched in Fig. 9.11) by a global plane subtraction (after image acquisition) in software.

The tilt can also be removed in real time using a tilt compensation features included in the control electronics. This is accomplished by orienting the *fast* scan direction so it is perpendicular to the maximum tilt direction. Figure 9.11(a) shows the situation when the fast scan direction is oriented parallel to the direction of maximum tilt. The drawback of this alignment is that the feedback circuit must rapidly adjust the z-positioner during the course of a given line scan in the fast scan direction. By orienting the fast scan direction as shown in Fig. 9.11(b), the rapid variation in the z-positioner is minimized since now the substrate tilt is located along the slow scan axis. The result of this preferred scanning alignment is usually a more stable image.

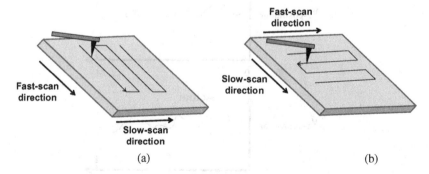

(a)   (b)

Fig. 9.11   When obtaining images with an AFM, it is usually important to orient the fast scan direction so it experiences a minimum change in z-height over one scan line. In (a), the fast scan direction is oriented so that the fast scan direction follows the incline of the substrate. In (b), the scan direction has been rotated using Eq. (9.37). The outcome of this realignment in the fast scan direction is usually a more stable image. The distances between scan lines are greatly exaggerated to make the point.

9.6 The Feedback Loop

A proper adjustment of the feedback controller is important for the operation of any AFM. It is important to understand that an AFM topographic image is derived from plotting the time varying voltage to the z-positioner of the substrate. In contact mode AFM, the time varying voltage is controlled by introducing an additional constraint that requires the signal from the photodiode to remain constant with time. The situation is sketched in Fig. 9.12 which shows the tip scanning once across a substrate. The length of the scan is L and it is assumed that the line scan requires a time Δt as discussed in Fig. 9.10. If the substrate's position can be adjusted with high precision at a rapid rate, the voltage to the z-positioner plotted in Fig. 9.12 to maintain a constant photodiode voltage will accurately track topography as indicated. In reality, the substrate's position cannot be adjusted with sufficient speed or accuracy to achieve this ideal condition, so the voltage to the z-positioner will only approximate the topography.

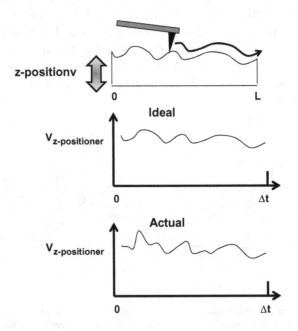

Fig. 9.12 The feedback loop controls how accurately the cantilever motion tracks the topography of the substrate. If the feedback loop responds with sufficient speed and accuracy, a faithful image of the substrate will be obtained. If the feedback loop is improperly set, then errors in the topography will result.

Understanding the feedback loop operation is critical to achieving scans that are close to the ideal. As initially introduced in Chapter 7, the AFM feedback loop usually contains two gain settings, K_P and K_I, to control the proportional and integral functions of the feedback. Each commercial AFM system uses a different method to set the values of K_P and K_I. The optimal feedback settings for one system have little to do with the optimal values for another system.

The frequency response required for the feedback to be effective is dictated by the time variation of the signal plotted in Fig. 9.12. The bandwidth for effective operation is related to (i) the total time (Δt) required to scan one line and (ii) the size of the features (especially sharp discontinuities) in the substrate that the tip must track. In general, the proportional (P) controller allows the probe to follow smaller, high frequency features by adjusting K_p. The backward integrator (I) controller allows the probe to follow larger, low frequency features by adjusting K_I.

The frequency response of the controller should extend by a factor of about 10 beyond the highest frequency required to reconstruct the sharpest feature that is measured. If the substrate has very abrupt features, then even though a typical scan frequency is low (typically 1 Hz), the controller must have a frequency response that extends well into the 10's of KHz to respond fast enough to accurately track that one feature. If feedback is too slow, blurred images will result since the sharp features will be inaccurately rendered. A feedback that responds too slowly is indicated by repeated tip crashes. If the feedback is too fast, feedback oscillations or overshoots will result as the tip tries to react too rapidly to noise impulses in the system. What determines the optimal feedback settings? The answer is the user. Feedback adjustment tends to be a creative process, with no analytically determinable answer. Students who are proficient at scanning develop an ability to make the proper adjustments in a judicious and time efficient way.

To better understand the nature of the problem, it is worthwhile to consider the characteristic response of a feedback loop to an abrupt step-like change in height. Figure 9.13(a) shows a step change at $t = 0$ which serves as the input to a feedback loop. Figure 9.13(b) illustrates schematically the generic output of the feedback loop in response to the step function input. The response displays a number of characteristic features such as a delay time T_L, a rise time T_R, an overshoot, an oscillation about the desired value, and a settle time in which the response finally enters into an acceptable error band about the desired output. A properly adjusted

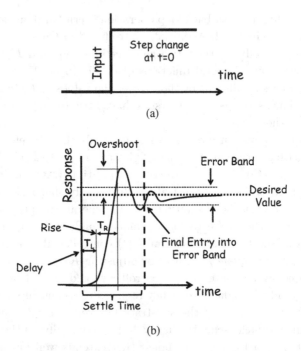

(a)

(b)

Fig. 9.13   The response of a generic feedback loop to a discontinuous step-function change at the input. In (b) the response of the feedback loop requires time to track the abrupt change. Until the desired output is reached, the loop will exhibit time delay, a rapid rise followed by an overshoot, and an oscillatory settling behavior.

feedback loop will have a minimum delay time, a rapid rise time, a minimal overshoot and a fast settle time.

Many tuning rules have been proposed to optimize the performance of a feedback loop. Perhaps the most famous, if not the first, is the Ziegler-Nichols rule that was introduced in 1942 [ziegler42]. The Ziegler-Nichols tuning rule is designed to endow feedback loops the best disturbance rejection by requiring aggressive gain and overshoot. Other criteria or rules must be established if one desires to minimize or eliminate (say) overshoot.

The Ziegler-Nichols criteria for determining K_p and K_I require estimates for the delay and rise times denoted by T_L and T_R in Fig. 9.13(b) to a step change. Once these times are known, Table 9.3 lists the predictions for the best values of K_P and K_I.

The problem with implementing such a definite rule for an AFM feedback loop is that once a cantilever and substrate are mounted, it

Table 9.3 The Ziegler-Nichols criteria for adjusting feedback gains. The gains K_P and K_I are set based on the measured rise time (T_R) and lag time (T_L) defined in Fig. 9.13.

Controller	K_P	K_I
P only	T_R/T_L	0
P & I	$0.9T_R/T_L$	$0.27T_R/T_L^2$

is difficult to perform controlled experiments to measure such qunatities as T_L and T_R. A further consideration is that specific tuning rules like Ziegler-Nichols are often designed to give an acceptable disturbance rejection performance for a specific application. If the application is not consistent with good dynamic tracking in an AFM, the rule may not be particularly relevant.

When acquiring AFM images, overly aggressive gains tend to make the system respond too quickly, producing erratic behavior. In contrast, simple assumptions based on pure time delays may not match reality. Experience teaches that while it is easy to have in mind a specific tuning rule, it is difficult to know whether it will be effective for a particular application. What is even worse, you might not realize that your preconceived tuning rule is inappropriate for the particular combination of tip, substrate, and scanning speed under study.

To make progress, it becomes important to develop an appreciation for what the parameters K_P and K_I control. Increasing K_P tends to reduce the rise time, increase the overshoot, has little effect on the settling time, and decreases the steady state error. Increasing K_I tends to decrease the rise time, increase the overshoot, increase the settling time, and reduces steady state error.

An empirical procedure for adjusting K_P and K_I to their optimal values in an AFM experiment might include the sequence of steps given below:

1. Select a scan speed and begin to acquire an image.
2. Set the integral gain close to zero ($K_I = 0$)
3. Systematically increase K_P until the tip first begins to oscillate. This will appear as grainy nose in the image. Then reduce K_p to $\sim 2/3$ the value at which oscillations were first observed.
4. Increase the integral gain until oscillations reappear.
5. Reduce K_I in steps until the "grainy" noise disappears across the image.

6. Consider reducing the set point force. Lower loading forces will minimize probe tip damage!
7. The feedback gains may need readjustment during an AFM experiment, since *set point, scan speed and feedback gain are all interrelated.*
8. If scanning is too slow, increase scan speed and repeat procedure starting at step 2.
9. Finally, analyze the error map image. When the error map is minimal, then the feedback controls are usually optimized.

9.7   Chapter Nine Summary

This chapter discussed the various calibrations that are often required to obtain meaningful AFM data. While modern AFM systems often have integrated applications to make the various calibrations easy to perform, it is useful to understand at some level what these applications entail. The different techniques devised to calibrate the flexural spring constant of a thin rectangular cantilever were reviewed and the calibrations of the X, Y and Z scanners were discussed. The standard procedure for calibrating the sensitivity of the AFM photodiode was presented. A discussion of the parameters required to estimate image acquisition times was given. Lastly, the generic feedback loop was reviewed and general tuning rules that optimize the AFM feedback loop were outlined.

9.8   Further Reading

References to the Original Literature:

[baird95]   D.C. Baird, Experimentation: *An Introduction to Measurement Theory and Experiment Design*, 3rd edition, Prentice-Hall, NJ (1995).

[butt95]   H-J. Butt and M. Jaschke, "Calculation of thermal noise in atomic force microscopy", Nanotechnology **6**, 1 (1995).

[cleveland95]   J. P. Cleveland, S. Manne, D. Bocek and P.K. Hansma, A non-destructive method for determining the spring constant of cantilevers for scanning force microscopy, Rev. Sci. Instrum. 64, 403 (1995).

[cook06]   S.M. Cook, T.E. Schaffer, K.M. Chynowth, M. Wigton, R.W. Simmonds, and K.M. Lang, "Practical implementation

of dynamic methods for measuring atomic force microscope cantilever spring constants", Nanotechnology **17**, 2135–2145 (2006).

[eaton01]   *Atomic Force Microscopy*, P. Eaton and P. West, Oxford University Press, Oxford, UK (2010).

[gates07]   R. S. Gates and M. G. Reitsma, "Precise atomic force microscope cantilever spring constant calibration using a reference cantilever array", Rev. Sci. Instrum. **78**, 086101 (2007).

[hutter93]   J.L. Hutter and J. Bechhoefer, "Calibration of atomic-force microscope tips", Rev. Sci. Instrum. **64**, 1868 (1993).

[hydrodamping]   The application and on-line calculator can be found at http://www.ampc.ms.unimelb.edu.au/afm/calibration. html# normal (last accessed June 23, 2013).

[levy02]   R. Levy and M. Maaloum, "Measuring the spring constant of atomic force microscope cantilevers: thermal fluctuations and other methods", Nanotechnology **13**, 33–37 (2002).

[sader98]   J.E. Sader, "Frequency response of cantilever beams immersed in viscous fluids with applications to the atomic force microscope", J. Appl. Phys. **84**, 64 (1998).

[sader99]   J. E. Sader, J. W. M. Chon and P. Mulvaney, "Calibration of rectangular atomic force microscope cantilevers", Rev. Sci. Instrum. 70, 3967 (1999).

[squires01]   G.L. Squires, *Practical Physics*, 4th edition (Cambridge University Press, Cambridge, UK, 2001).

[te reit11]   J. te Reit, *et al.*, "Inter-laboratory round robin on cantilever calibration for AFM force spectroscopy", Ultramicroscopy **111**, 1659–1669 (2011).

[wagner11]   R. Wagner, R. Moon, J. Pratt, G. Shaw and A. Raman, "Uncertainty quantification in nanomechanical measurements using the atomic force microscope", Nanotechnology **22**, 455703–11 (2011).

[ziegler42]   J.G. Ziegler and N.B. Nichols, "Optimum settings for automatic controllers", Trans. ASME **64**, 759–768 ((1942).

Further reading

[bragg74]   G.M. Bragg, *Principles of Experimentation and Measurement*, Prentice-Hall, NJ (1974).

[butt05]   H.-J. Butt, B. Cappella, M. Kappl, "Force measurements with the atomic force microscope: Technique, interpretation and applications. Surface Science Reports **59**, 1–152 (2005).

[beckhhoefer05] J. Bechhoefer, "Feedback for physicists: A tutorial on control", Rev. Mod. Phys. **77**, 783–836 (2005).

9.9   Problems

1. The spring constant of the first mode of a rectangular cantilever is calibrated using the thermal fluctuation method. A measurement of the power spectral density (PSD) of the cantilever provides the following data as a function of the frequency f:

Point No.	Frequency (Hz)	PSD (units of m^2/Hz)
1	63700	5.85E-25
2	63800	6.18E-25
3	63900	6.68E-25
4	64000	7.52E-25
5	64100	7.85E-25
6	64200	8.18E-25
7	64300	1.25E-24
8	64400	1.59E-24
9	64500	3.01E-24
10	64600	4.34E-24
11	64700	2.84E-24
12	64800	1.59E-24
13	64900	1.02E-24
14	65000	8.02E-25
15	65100	7.85E-25
16	65200	6.68E-25
17	65300	5.85E-25
18	65400	6.01E-25
19	65500	5.68E-25
20	65600	5.68E-25

Using MATLAB or EXCEL, perform a rough fit to this data by adjusting the three fitting parameters (plus a constant Background) that appears in the fitting function for the PSD

$$\mathrm{PSD}(f) = \mathrm{Background} + \frac{f_o^4 A_1}{\left[(f_o^2 - f^2)^2 + \left(\frac{f_o^2}{Q^2}\right) f^2\right]}$$

where A_1 is a constant, f_o is the resonant frequency of the cantilever, and Q is the quality factor. Once values for these parameters are estimated, calculate the spring constant of the cantilever using Eq. (9.12)

$$k_1 = \frac{2k_BT}{\pi A_1 Q f_o}.$$

Assume the data was taken in a laboratory with a temperature of 20°C.

If the fitting is done properly, a value for k_1 near $2\,\text{N/m}$ should result. In a careful experiment, more than 20 data points listed in the table would be used to characterize the PSD and a least squares fit would be implemented to obtain estimates for the fitting parameters.

2. Suppose the cantilever measured in Problem 1 above is etched from a Si wafer having a [100] orientation. A careful measurement of its length and width give values of $250\,\mu\text{m}$ and $35\,\mu\text{m}$, respectively. What must be its effective thickness to match the value of the spring constant obtained in Problem 1?

3. Using the values of f_o and Q found in Problem 1 and the dimensions of the cantilever given in Problem 2, evaluate the spring constant using Sader's hydrodynamic theory given by Eq. (9.18)

$$k_1 = 0.1906\,\rho w^2 L Q \omega_o^2 \Gamma_i(\omega_o)$$

The hydrodynamic factor Γ_i found in this equation can be conveniently evaluated by consulting the web site: http://www.ampc.ms.unimelb.edu.au/afm/calibration.html#normal

Assume the measurements were done in air at the standard atmospheric pressure of 1.01×10^5 Pa and a temperature of 20°C. The viscosity of air under these conditions is about $1.8 \times 10^{-5}\,\text{kg}\,\text{m}^{-1}\text{s}^{-1}$. To successfully complete this problem, an updated version of Java must be installed and working on your computer.

How does the answer for k_1 using the hydrodynamic model compare to the value of k_1 determined using the thermal tuning method?

4. A plot of typical data from an AFM deflection vs. displacement experiment is given below. Suppose this data were taken by indenting a hard tip into a hard substrate.

(a) From an examination of the data, estimate roughly the optical sensitivity of the position sensitive detector.

(b) If a cantilever with a spring constant of $0.75\,\text{N/m}$ was used when the data were acquired, estimate the set point force in nanoNewtons.

(c) From an examination of the data, estimate the pull-off force in nanoNewtons.

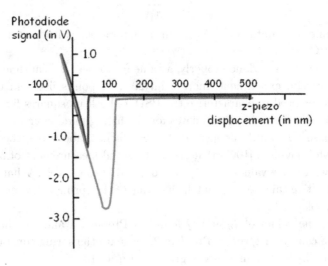

Chapter 10

Computer-Aided AFM Simulations

10.1 Introductory Comments

The previous chapters have described various aspects of Atomic Force Microscopy and many equations have been derived that are of general use. In this chapter we introduce the web-based software called VEDA (Virtual Environment for Dynamic AFM) for simulating a variety of AFM experiments. The software was developed by A. Raman and his students at Purdue University to facilitate a theoretical simulation environment for AFMs [melcher08], [kiracofe12], [nanohub13]. VEDA is free and open source and only requires that you have registered for a free account at the nano-HUB web site. The software requires no computer resources at the users end other than a web-browser equipped with an up-to-date version of Java. All computations are done remotely and the final results are displayed via the web on the user's browser. As of this writing, VEDA has been in use for over six years and well over 1500 users throughout the world have run AFM simulations using it.

VEDA is supported by the nanoHUB website which supports an array of computational tools with a focus on nanotechnology, class room courses, seminars and educational animations [nanohub_intro]. The underlying nanoHUB platform is quite flexible and supports numerous nano-related tools beyond VEDA that allow you to customize your web page for ease of use.

The VEDA software captures all aspects of cantilever motion and is designed for both beginners and advanced users alike. The simulation environment is especially useful for a number of reasons. First, VEDA simulations allow you to ask and answer "confidence building questions" that sharpen your understanding of AFM operation. Secondly, the simulations

are useful because they allow you to correlate AFM experiments with theoretical parameters required to interpret the data. This allows an AFM
user to better understand the relevant theory required to fit experimental
results. All too often, little or no effort is made to understand the underlying theoretical framework. Thirdly, the simulations are especially useful to
design meaningful AFM experiments in advance, before any experiments are
performed. By theoretically investigating a set of experimental conditions,
the optimal range for conducting an experiment can usually be identified.
Lastly, a VEDA simulation is a great way to improve your understanding
of data already published. By performing a VEDA simulation to reproduce published results, your understanding of prior work can be greatly
enhanced.

In what follows, we describe how to access the VEDA software and we
present a few simple examples that are relevant to contact mode AFM. The
simulation of dynamic mode AFM experiments will be discussed further in
Part II of the lecture notes.

10.2   Getting Started with VEDA

To access VEDA, you must first create an account (it's free) on nanoHUB
at http://nanohub.org. If you navigate to this site, you will find a banner
panel with a query window labeled **Login** and **Register**. If you already
have a login account, you can click on the **Login** query box and provide
your password. If you need to register with nanoHUB, click on the **Register**
query box and answer the questions that are asked. There is no charge for
opening a new account. After registering, you will receive an email with a
password that will allow you to access the extensive nanoHUB content.

After logging into nanoHUB, you can access VEDA by selecting the
Resources tab and then selecting **Tools** from the drop-down menu (See
Fig. 10.1(a)). By selecting **Tools** you will open up a list that is alphabet-
ically arranged listing all simulation tools available in VEDA. Locate the
listing **atomic force microscopy** (See Fig. 10.1(b)) under the column
heading **Tag** and click on it to see those tools related to AFM. A new list
will appear in a middle column.

In this list, click on the item marked **VEDA** and a window in a third column will open up showing the latest version of VEDA. You should see a box
labeled **Launch Tool**. By clicking this box, the VEDA tool should appear
in your web browser (Fig. 10.2) and you can now initiate AFM simulations.

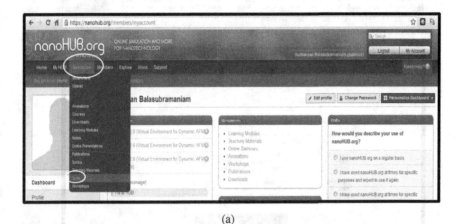

(a)

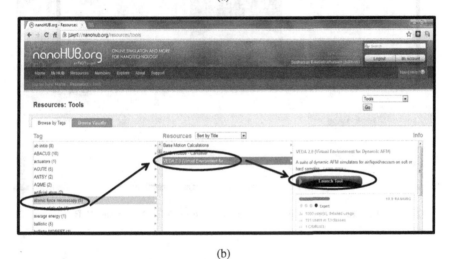

(b)

Fig. 10.1 Screenshots of relevant web pages at nanoHUB.org. In (a), a screen shot of the main nanoHUB web page. Select **Resources**, then **Tools**. After **Tools** has been selected, a new web page will open as shown in (b). Under the column labelled **Tag**, select **atomic force microscopy**, then **VEDA**.

The VEDA homepage should appear as in Fig. 10.2 and it reveals a number of useful features, including the ability to share a simulation session with other registered users of nanoHUB. For now, the most relevant feature is the **Pick a tool here** selection box. When selected, a drop down menu will list all the tools that have been developed within VEDA (see Fig. 10.3).

Fig. 10.2   Screenshot of the homepage for VEDA on nanoHUB.

The list is extensive and includes ~15 different tools that allow you to accurately run a variety of different AFM simulations.

Before investigating the tools further, it is worthwhile to provide a high-level overview of how best to use VEDA. This overview is provided in Fig. 10.4 and indicates that VEDA is especially useful for running simulations where a **few** parameters are systematically varied to learn what influence they might have on an AFM experiment. This feature is indicated by the loop formed between the initial "Select Simulation Parameters" box and the "Output" box which allows you the possibility to "Change Input Parameters".

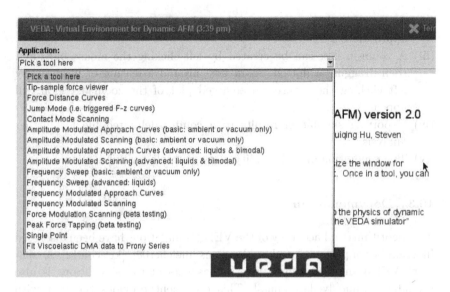

Fig. 10.3 The AFM simulation tools available in VEDA.

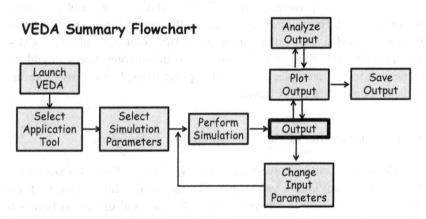

Fig. 10.4 High level view of how VEDA might be best used. The VEDA simulation software is designed to allow you to investigate how changes to a few input parameters affect the simulated output. All motions of the cantilever are faithfully captured by the simulation software.

It should be realized that VEDA allows the "Output" of a simulation to be further manipulated in a variety of ways that include

(i) A precise readout of any (x, y) value on a graph by positioning the mouse cursor on the plotted result.

(ii) A change in the x or y axis scale by double clicking the mouse on the appropriate axis.

(iii) A zoom-in box can be created by left-clicking the mouse once and then dragging a dotted 'zoom box' around a feature of interest. By left clicking the mouse, an enlarged plot of the zoomed region will appear.

(iv) Exporting simulation results to a comma-delimited file for further analysis

(v) The generation of high-quality plots, suitable for publication.

10.3　Documentation

The algorithms and accuracy of the VEDA simulations have been discussed in a number of peer-reviewed publications [melcher08], [kiracofe12]. A complete VEDA manual summarizing all aspects of the simulations is also available on-line [veda_manual]. This document provides the user with the necessary information about the theory, modeling, and computational approach that are present in the VEDA simulation tools, and the inputs required to perform simulations. Additionally, several demonstration examples are provided that point out some of the nonlinear phenomena that underpin the operation of the atomic force microscope. Users should use these well-tested examples as a starting point to explore the vast parameter space present in AFM [caveat].

10.4　Simulations of the Tip–Substrate Interaction

In this section, we will investigate the *Tip–Sample Force Viewer* tool in VEDA. This tool allows you to plot the tip–sample interaction that was discussed extensively in Chapters 4 and 5. The goal of this section is to illustrate how one might compare the Hertz and DMT models for tip-sample interaction.

After selecting the *Tip–sample force viewer* tool in VEDA, a new window will open with three separate tabs labeled ❶ *Operating conditions*, ❷ *Tip–sample interaction properties*, and ❸ *Simulate*. Only one tab can be selected at a time and the parameters required to perform the simulation must be entered into the input boxes listed when each tab is open. You can click on each numbered tab in any order to navigate between them. Usually, default parameters are listed when a tab is first opened. These default parameters have been carefully chosen to provide a guide to what might be reasonable input values for VEDA. When starting to use a new tool, it is useful to initially select the default parameters in order to first obtain a meaningful simulation result.

When the tab ❶ *Operating conditions* is selected, the user must input values for the initial tip–sample gap, the final gap separation, the number of points plotted, and the tip velocity. For example, if the initial tip gap is set to 3 nm and the final gap separation is set to -1 nm, then the simulation will be run for gap values from $+3$ nm to -1 nm. If you desire to reverse the direction of the tip and run a simulation from -1 nm to $+3$ nm, then these values must be entered instead. Usually, selecting a simulation with 1000 data points is reasonable. For this simulation, we do not require any special behavior regarding the tip velocity, so selecting the default 'quasi-static' option is recommended.

When the tab ❷ *Tip–sample interaction properties* is selected, a new window should appear that will allow you to select the tip–sample interaction model from a drop-down menu. Let's first plot the predictions of the Hertz model, so select "Hertz contact" from the drop-down menu. When this model is selected, a representative schematic diagram is displayed which should remind you of the essential features of the Hertz model discussed in Chapter 5. Below this graph, the user must input values for a number of important parameters. Sometimes, if your browser window happens to be too small, these parameters may not be visible. If this happens to be the case, expand the size of your window and you should see a scroll bar appear on the far right side of the window along with a few boxes that require input parameters. Not all of these boxes may be visible on the screen, so you may need to scroll down using the slider bar to enter simulation values.

As shown in Fig. 10.5, VEDA requires you to enter various parameters in order to complete the simulation. This allows you to systematically investigate the effects of various parameters on AFM experiments. Note that some of the windows may appear in light gray which indicates the values are inaccessible for the options that you have selected. So for example in Fig. 10.5, once a "Classical sphere" has been selected for the tip shape, the cone half-angle parameter becomes inactive. To activate the cone half-angle option, the "Cone option" in the Tip shape window must first be selected.

If all the default parameters listed are selected, VEDA is now in a position to calculate the tip–sample interaction force as a function of position between 3 nm and -1 nm using the Hertz model for a tip of radius 10 nm with a Young's modulus of 130 GPa and a Poisson's ratio of 0.3

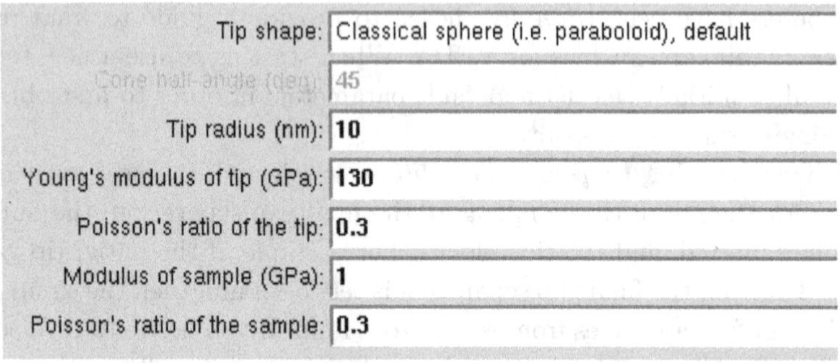

Tip shape:	Classical sphere (i.e. paraboloid), default
Cone half-angle (deg):	45
Tip radius (nm):	**10**
Young's modulus of tip (GPa):	**130**
Poisson's ratio of the tip:	**0.3**
Modulus of sample (GPa):	**1**
Poisson's ratio of the sample:	**0.3**

Fig. 10.5   Input parameters required to run a Hertz contact model between a tip and a substrate.

indenting into a sample with a Young's modulus of 1 GPa and a Poisson's ratio of 0.3.

Note that a variety of advanced options may also appear. Usually a reference to the original literature is provided if you wish to investigate these options further. For our purposes, we will not select any of these advanced options here.

The simulation can be run by clicking on the ❸ *Simulate* tab on the top of the page or alternatively, by clicking on the Simulate button that can be found on the bottom right of the ❷ *Tip–sample interaction properties* window.

After initiating the simulation, a new window should appear in a few seconds with a plot the Hertz tip–sample interaction model for the parameters that you have selected.

In order to compare to the Hertz model to the tip–sample interaction model using DMT, you must next select DMT by clicking the mouse pointer in the *Tip–sample interaction* window. When you do this, a drop down menu will appear. Once again, a representative schematic diagram of the DMT model will be displayed along with a window of parameter boxes (Fig. 10.6) that must be completed.

To make a consistent comparison, it might be a good idea to use the same tip shape, tip modulus, sample modulus, and Poisson ratios selected for the Hertz simulation. Because the DMT model includes long-range interactions, a Hamaker constant must be entered to indicate the strength of this interaction. There are a number of possible options that can be selected to specify this tip–sample interaction. In this example, we choose to enter in the interatomic distance (0.3 nm) and a Hamaker constant (3×10^{-19} J)

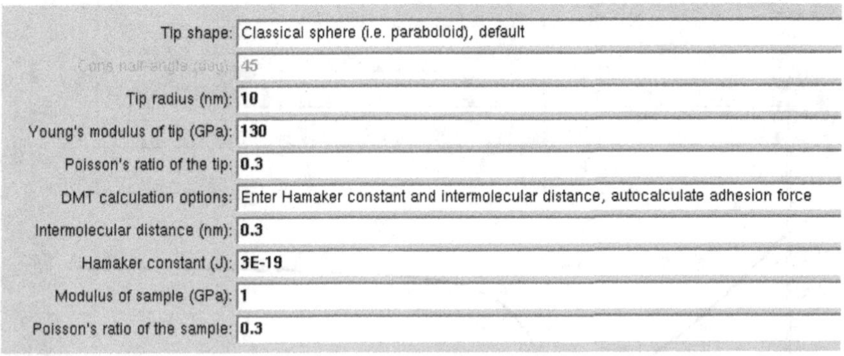

Tip shape:	Classical sphere (i.e. paraboloid), default
Cons half-angle (deg):	45
Tip radius (nm):	10
Young's modulus of tip (GPa):	130
Poisson's ratio of the tip:	0.3
DMT calculation options:	Enter Hamaker constant and intermolecular distance, autocalculate adhesion force
Intermolecular distance (nm):	0.3
Hamaker constant (J):	3E-19
Modulus of sample (GPa):	1
Poisson's ratio of the sample:	0.3

Fig. 10.6   Input parameters required to run a DMT contact model between a tip and a substrate.

Fig. 10.7   The Results window controls which simulation is plotted. In this screen shot, the window indicates two simulations (2 results) have been performed. The horizontal plot bar allows you to select which plot is displayed by adjusting the position of the pointed arrow. By clicking on the *All* button, all the simulations can be displayed at once.

and let VEDA calculate the appropriate adhesion force. The parameters chosen are indicated in Fig. 10.6.

By clicking on the ❸ *Simulate* tab, a plot of the DMT model will now appear after a few seconds. Note that the parameters already selected under the tab ❶ *Operating conditions* (e.g., the initial tip gap and the final separation) remain the same.

Which plot is displayed — Hertz or DMT — can be selected using the plot bar that appears in Fig. 10.7. Careful examination of this plot bar indicates that two simulations have been run. A plot of each simulation can be displayed by clicking on the small vertical bars that appear in the plot bar window.

When the small vertical line (cursor) is selected in the plot bar, it becomes bold and a small pointed arrow appears beneath it. At the same time, the title over the plot bar displays a descriptive comment to remind you what simulation you have selected to view.

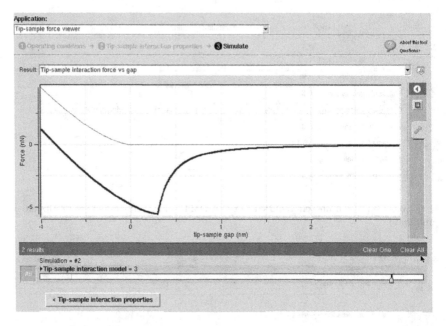

Fig. 10.8   A comparison of the Hertz and DMT contact models for a 10 nm radius tip with an elastic modulus of 130 GPa interacting with a sample having an elastic modulus of 1 GPa.

If all you wish to view all simulations performed during a session, click the *All* button that appears to the left of the plot bar. If this is done for the simulations under discussion, you should see a plot similar to that shown in Fig. 10.8. This plot compares the predictions of both the Hertz and DMT tip–sample interaction model for the parameters chosen.

Since Hertz neglects long-range van der Waals interactions, the interaction force between the tip and sample remains zero as the tip-sample gap is decreased. The DMT model includes the long range van der Waals interactions and hence the tip experiences a negative (attractive) force as it approaches the sample. The magnitude of the attractive force is dictated by your choice of the Hamaker constant and the interatomic distance which you selected.

If you wish to change the range over which the simulation is plotted, you can do so by positioning the mouse cursor over the x or y axis and double clicking with the mouse. A window will open that allows you to input the exact axis limits to the range you desire.

You can zoom into a region of the graph and replot it by positioning the mouse cursor on a selected region of the plot, left clicking the mouse, and dragging a dotted box that includes the region you wish to replot. A final left click of the mouse will generate a zoomed-in graph of the region you have selected.

The box labeled *Result* in the plot window above the graph allows you to select a variety of different plots that are all calculated for a specific VEDA simulation. For the *Tip–Sample Force Viewer* tool, the default plot is just the *Tip–sample interaction force vs. gap*. By clicking on the *Result* box, other plots can readily become accessible.

A useful feature found in the *Result* box is the *Echo of input parameters*. By selecting this feature, you will see a list of all the input parameters used in a particular VEDA simulation. By saving such a list, you can always reproduce a simulation at some later time.

Additional simulations using different input parameters can always be added at any time by selecting the ❷ *Tip–sample interaction properties* tab or clicking on the box located in the lower left of the plot window which is labeled *Tip–sample interaction properties.*

The three icons that appear along the right-edge of the plot displayed in Fig. 10.8 provide access to a variety of useful tools.

If the *"About this tool"* box next to the green question mark is selected, a window will appear that gives background information about the VEDA tool. If the *"Questions?"* box next to the green question mark is selected, you can view some of the recent user questions that have been asked and the answers that have been provided. Selecting this option also allows you to ask a question.

If the *downward green arrow* icon is selected, a variety of options will appear that allow you to export the numerical results of a calculation to an external file or to produce a high quality plot.

If the *leftward pointing blue arrow* icon is selected, a window will open that will allow you to change various aspects of the plot displayed.

If the *square box* icon is selected, you can quickly reset the plot to its original appearance.

A helpful pop-up window briefly describing the function of each icon can be viewed by mousing over any icon with the mouse pointer.

When ending a VEDA simulation, it is advisable to click the × *Terminate* box that appears in the upper right of the main VEDA window. Alternatively, a → *Keep for later* box is also available. Since VEDA runs on the nanoHUB platform, you should realize that only three jobs can be saved before the nanoHUB platform will object. This feature prevents users from inadvertently opening an unreasonably large number of simulations.

10.5   Simulations of Force-Distance Curves

The *Tip–Sample Force Viewer* tool is fast and very straightforward and serves as a useful introduction to how VEDA operates. In this section, we will discuss the *Force Distance Curves* tool which is designed to allow you to simulate the tip–sample force as a function of tip–sample separation.

After selecting the *Force Distance Curves* tool, a new window will open up with four separate tabs labeled ❶ *Operating conditions and cantilever properties*, ❷ *Tip–sample interaction properties*, ❸ *Simulation parameters*, and ❹ *Simulate*. In the discussion that follows, we will rely on a pre-loaded example to illustrate the important features of this tool. Start by clicking on the *Example loader* box and select the *FZ curves Example 2: Approaching and retracting from a sample modeled using JKR contact* that is found under the ❶ *Operating conditions and cantilever properties* tab. This is a non-trivial example of the force vs. distance curves calculated using the JKR contact model discussed in Chapter 5.

The default parameters for this simulation now include the spring constant of the cantilever ($0.87\,\mathrm{N/m}$), the quality factor for the cantilever ($Q = 33$) and the natural resonant frequency of the cantilever ($f_o = 44\,\mathrm{kHz}$). For simplicity, the cantilever is allowed to oscillate in only one eigenmode specified by the $\alpha = 1.8751$ parameter given by Eq. (6.39) and Table 6.1. In principle, you could enter other values for these parameters to match experiments you may be conducting instead of these default values. To perform a realistic simulation, the tip must approach and then retract from the sample. To insure this happens, select the *Approach and retract* option when the mouse pointer is clicked inside the *Choose operating mode box*.

If the ❷ *Tip–sample interaction properties* tab is selected, a representative plot of the JKR tip–substrate model is displayed and a number of input boxes will appear below. These input boxes already have preloaded values for this particular simulation. The tip shape required by the JKR model is by default a sphere and this shape cannot be changed. The JKR model is applicable for a soft sample, and in this example, Young's modulus of the sample is 1 GPa, while the tip modulus is set to 130 GPa. The tip radius is selected to be $10\,\mathrm{nm}$.

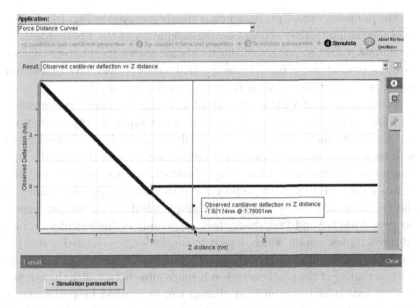

Fig. 10.9 A screen capture of a VEDA simulation for a tip that approaches and retracts from a sample. The tip–sample interaction is specified by the JKR model.

Under the ❸ *Simulation parameters* tab, you can adjust parameters that control the accuracy and speed of the simulation.

Finally, by clicking the ❹ *Simulate* tab, the VEDA simulation will begin. The progress of the calculation can be observed by viewing the *Performing simulation* task bar. A default graph (see Fig. 10.9) will be displayed which shows the observed cantilever deflection q (nm) as a function of the z-displacement (nm) when the tip approaches and retracts from the sample over the range ($+10$ nm to -5 nm) specified under the ❶ *Operating conditions and cantilever properties* tab.

Figure 10.9 illustrates the approach and retract curves and shows how a cursor, positioned over any part of the curve can be used to read out the simulation results with high accuracy. For the case shown in Fig. 10.9, the cantilever snaps out of contact with the sample at a z distance of 1.78 nm. The cantilever deflection at this point is -1.62 nm (the minus sign implies the cantilever is bent toward the sample). Since the cantilever spring constant is $k = 0.87$ N/m, the adhesive force is about 1.41 nN when the cantilever snaps from the substrate. This value of course was set by the value appearing in the *van der Waals Adhesion force* box which can be found in the ❷ *Tip–sample interaction properties* tab.

The jump-to-contact phenomenon is clearly observed in Fig. 10.9 by the small discontinuity that appears during simulation of the tip approaching the sample. Close inspection of the simulation shows the jump-to-contact feature is plotted at $z = 0$ nm. Since the JKR model is based on a continuum model, the interatomic spacing parameter a_o is not defined.

The real power of VEDA is revealed by clicking the mouse cursor inside the *Result* box and observing the many different types of outputs that are now readily available. Not only can the observed cantilever deflection (q) vs. z be plotted, but the Tip-sample interaction force (kq) vs. z, the tip–sample gap (d) vs. z, the Tip-sample interaction force vs. gap, and the sample indentation vs. z. Inspection of the indentation plot vs. z indicates that the tip penetrates ~ 1 nm into the sample when the tip displacement z has gone 5 nm beyond jump-to-contact. A zoom-in of the indentation plot will reveal a few points which appear to be randomly scattered about the main simulation curve. Although these points appear to be noise, they are due to a ringing of the cantilever when it snaps into contact with the sample. The full extent of this cantilever ringing can be verified by performing a second simulation which spans a much narrower z-range (say from $+1$ nm to -1 nm).

10.6   Simulations of Contact Mode AFM Scans — Influence of Feedback Gain

VEDA simulates a topographical image that an AFM would measure. To perform this function, VEDA has a number of well-characterized built-in structures called *features* that are user selectable. These structures include a step, a trapezoidal plateau or trench, a smoothly varying sinusoidally-shaped hill or crater, and a cylinder. It is possible to specify different elastic parameters for a feature and the sample on which the feature is supported. Using this capability, scans of objects deposited onto flat substrates can be investigated. Also, the topography simulation now requires a feedback loop, requiring the user to specify the gains of the P and I controllers. Furthermore, the simulation tracks many useful variables, allowing plots of such quantities as the peak force exerted by the cantilever vs. position, a quantity that is difficult to know when performing an AFM experiment.

To explore this capability, you first need to select the *Contact Mode Scanning* tool. A new window will open up with five separate tabs labeled ❶ *Operating conditions + Cantilever properties,* ❷ *Tip–sample interaction properties: substrate,* ❸ *Simulation parameters,* ❹ *Tip–sample interaction properties: Feature,* and ❺ *Simulate.* As before, the parameters required to perform the simulation must be entered into the input boxes listed when each tab is open. You can click on each numbered tab in any order to navigate between them. Automatic default parameters are listed when a tab is first opened. These default parameters have been carefully chosen to provide a guide to what values VEDA is expecting.

To begin, let's investigate the topography that is measured when a tip is scanned across a rectangular feature that has a length of 50 nm and a height of 30 nm.

Under the ❶ *Operating conditions* + *Cantilever properties* tab, select 1 eigenmode for the cantilever that has a spring constant of 3 N/m. Let's set the quality factor of the cantilever equal to 50, a value that is presumably measured when the calibration of the cantilever's spring constant is performed as described in Chapter 9. To be specific, specify a natural frequency of the cantilever of 60 kHz and a setpoint deflection of 2 nm. The radius of the tip will be set to 10 nm. We will set the scan lines per second to equal 10. We must also specify the Proportional and Integral feedback gains. To begin, we'll use 0.005 for both values. Although these absolute numbers will not directly apply to your AFM instrument, the relative changes to simulate an AFM scan will reflect any adjustments you make while performing an AFM experiment. It is also worth mentioning that these values are critical in determining whether a simulation will run successfully. If they are set too high or too low, the numerical feedback loop in VEDA will not converge properly. The simulation will time out or issue an error message to the user.

Under the ❶ *Tip–sample interaction properties: substrate* tab you can specify the Tip–sample interaction model. In what follows, let's select the Hertz contact model. We'll use a tip radius of 10 nm and a modulus/ Poisson ratio for the tip of 130 GPa/0.3. The modulus/ Poisson ratio for the sample will be modeled by using the parameters 100 GPa/0.3.

Under the ❸ *Simulation parameters* tab, use the default values given except for the scan size length which we will set to 500 nm.

The ❹ *Tip–sample interaction properties: Feature* tab allows you to select the shape of the feature supported on the substrate and specify its elastic properties. Here we will select a *Step* feature that is +30 nm high and 50 nm wide. (If the height is set to −30 nm, then a trench 30 nm deep is created). For this example, we'll simulate an AFM contact mode scan over this feature using the **Hertz** contact model. The modulus/ Poisson ratio for the feature will be modeled using the parameters 1 GPa/0.3. The feature is therefore considerably softer than the substrate and tip.

Initially, we will **not** include tip dilation effects, but we will specify the material properties. This means the simulations plotted will not be dilated by the tip diameter, an important effect that will be discussed in Sec. 10.7. The advantage of not including tip dilation is that the simulations run quickly, and all input parameters can be optimized before including a more lengthy simulation involving the tip shape.

By selecting the ❺ *Simulate* tab, VEDA will begin the simulation. VEDA must approach the tip to the sample before starting a simulated scan, and initially you should see a "Computing transients" message in the plot bar below the graph. If the feedback gains are not properly set, the simulation never progresses beyond the "computing transients" stage and no plot will be displayed. If this happens, you must readjust the values for P and I.

After the tip has successfully approached the sample, an actual line scan of the tip over the specified feature will be performed. This is indicated by a "Performing simulation. . ." message in the plot bar. After a few tens of seconds, a plot will appear under the *Result* box which should display by default the *Measured topography* is shown in Fig. 10.10.

The plot will have two simulated AFM profiles — one colored blue and one colored red. The red line in the VEDA plot indicates the size of the feature that was specified — a rectangular box-like feature that is 30 nm high and 50 nm wide. This feature should be centered in the scan window. In the VEDA plot, the blue curve represents the motion of the sample in an attempt to maintain the specified 2 nm setpoint deflection. This represents the simulated topography measured by the AFM for the input parameters that were chosen. For completeness, a copy of the echo input parameters list is also provided in Table 10.1.

The plot contains a wealth of information and shows the trajectory of the tip as is scans over the softer, box-like feature. The measured height of the feature for the cantilever and set point deflection specified is 28.9 nm, about 1.1 nm less than the feature height. The discrepancy is due to an indentation of the tip into the soft feature (caused by the specified 2 nm cantilever setpoint deflection times the 3 N/m spring constant = 6 nN) in combination with the Hertz model.

The plot also shows that the sample motion does not precisely follow the feature geometry as the tip is scanned at a rate of 10 scan lines/s. The feedback parameters are not optimized to accurately track the rapid

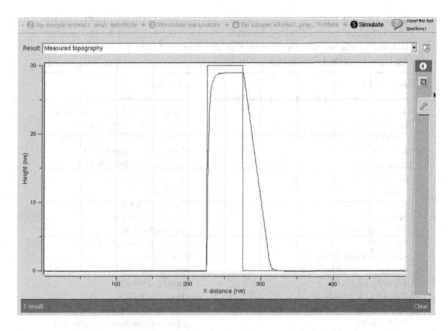

Fig. 10.10   VEDA output when a 30 nm high and 50 nm wide box-like feature is scanned in contact mode at constant force of 6 nN. The geometry of the rectangular feature is plotted along with a simulation of the AFM profile.

height variation of the leading and trailing step-like edges. The error on the trailing edge is particularly large and is due to the tip parachuting off the back edge of the feature. During this descent, the force on the tip is zero — in other words the tip is not in contact with the sample.

Figure 10.11 shows a zoomed in simulation in which the scan speed is systematically varied between 1 scan line/s to 10 scan lines/s. As is evident, the effect of the scanning speed on the measured topography is significant and clearly a scan speed of 10 scan lines/s is too fast for the feedback P and I values that have been chosen.

A variety of other plots can be displayed by clicking in the *Result* box. These plots include *Measurement error vs. x*, *Deflection error vs. x*, *Mean interaction forces vs. x*, and *Indentation vs. x*.

10.7   Simulations of Contact Mode AFM Scans — Influence of Tip Shape

The above simulations were performed without taking into account any geometric distortion due to the finite radius of the tip (in this case 10 nm).

Table 10.1 A list of input parameters used to generate the plot in Fig. 10.10. The values of many of the parameters listed are not required for the simulation under discussion. The parameters listed in bold have been explicitly set for the simulation shown in Fig. 10.10.

exc_choice = 1	want oscillatory F
operating mode 3 fexcite 3 **numModes**	want hydration F
1	Want_hsyteretic_hydration F
LockInTC (us)	electrostatic theta tip 10
0.000000000000000E+000	electrostatic theta lever 10
LockInOrder 0	electrostatic height 10
mtip 0	electrostatic length 100
omegad 9.3	electrostatic width 30
Ainitial = 1	VEchoice 1
omega = 60	WantCapAd F
Keq = 3.0	WantSurfHyst F
Chi = autocalc	**Esample 100**
Q = 50	**Poisson_sample 0.3**
sample_freq Mhz 1	KD 0.001
sample_freq 1000000.00000000	epsilon 80
LineSpeed 10	sigmat −0.0025
Z_feedback_choice 4	sigmas −0.032
SNratio 60	mat properties for:
KP 0.005	input.phase(feature).group(fp)
KI 0.005	want WLC F
HF 30	kts_R 10
LF 50	kts_A 10
LF2 30	Fadhesion 1.4167
LS 500	aDMT 0.2
Feature type 1WantTSCon F	A_hamaker 3.4E-20
tip shape =1	fts_model 2
tip angle 45	want tip squeeze F
Rtip 10	want oscillatory F
Etip = 130	want hydration F
Poisson_tip = 0.3	Want_hsyteretic_hydration F
mat properties for: input.phase(ts)	electrostatic theta tip 10
want WLC F	electrostatic theta lever 10
kts_R 10	electrostatic height 10
kts_A 10	electrostatic length 100
Fadhesion 1.4167	electrostatic width 30
aDMT 0.2	VEchoice 1
A_hamaker 3.4E-20	WantCapAd F
fts_model 2	WantSurfHystEs
want tip squeeze F	

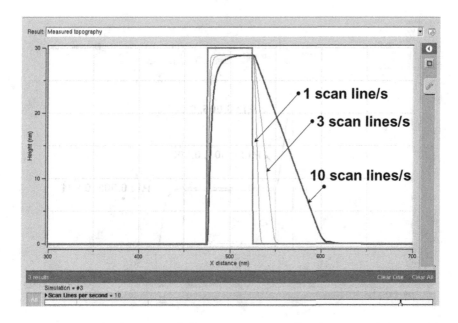

Fig. 10.11   The effect on AFM topography of different scan rates.

The tip shape can be taken into account by activating the *Include geometric convolution* box under the the ④ *Tip–sample interaction properties: Feature* tab. If this box is activated, the simulation now requires a few minutes to complete since the full geometry of the tip must now be included. After the simulation is complete, a plot similar to the bold solid line in Fig. 10.12 should be observed. This is the topographic scan that should be compared to an AFM experiment. The box-like feature, as specified in the ④ *Tip–sample interaction properties: Feature* tab, will be plotted in red.

It is clear from Fig. 10.12 that the topographic scan is dominated by a sharp spike-like characteristic which occurs at the leading step edge. Basically, the tip is overacting to the sudden cantilever deflection that occurs when the tip encounters the leading step edge. The relative values of the gain are set too high for the scan speed. To correct this anomaly, the P and I gains must be adjusted downward. The simulation loop illustrated in Fig. 10.4 is ideal for holding all parameters fixed while a few are systematically adjusted. The results of a few different iterations on P and I are also illustrated in Fig. 10.12 where it is found that the anomalous feature is

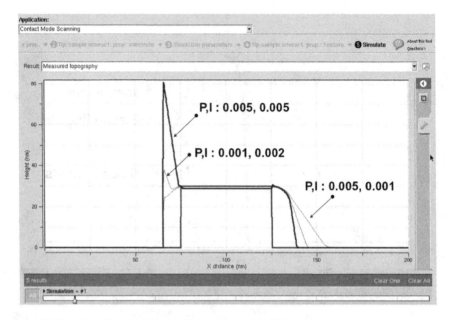

Fig. 10.12   The effect of different P, I settings on AFM topography. In this example, a finite tip radius of 10 nm has been included in the simulation.

most sensitive to the value of the I feedback gain. This example illustrates what course of action is required if you were performing an AFM experiment and observed such a rapidly increasing anomaly when scanning over a sharp feature.

The simulation also allows you to estimate the observed width of the feature and compare it to the known width. For example, a careful measurement of the width of the P, I: 0.001, 0.002 scan gives a full width at half-maximum of $138.6 - 55.2$ nm $= 83.4$ nm for the 50 nm wide step. Since the tip has a radius of 10 nm, we might expect the 50 nm wide feature to be broadened to 70 nm. The additional distortion of 13.4 nm is due to the inability of the feedback to accurately track the feature.

10.8   Chapter Ten Summary

An overview of the AFM simulation software VEDA (Virtual Environment for Dynamic AFM) has been provided. This AFM simulation software

Table 10.2   A template listing the required input parameters for a force-distance simulation in VEDA.

VEDA Worksheet: Force-Distance Curves

Simulation Notes: _____

Simulation Date: _____

"Operating cond. + Cantilever prop."		"Tip-sample interact. prop."	
Operating mode		Tip-sample interaction	
No. eigenmodes		R_{tip}	
Cantilever stiffness		E_{tip}	
Q		ν_{tip}	
Natural frequency		vdW adhesion force	
Z approach/retract speed		E_{sample}	
Gamma (z-drag)		ν_{sample}	
Initial separation			
Final separation			
"Simulation parameters"			
Num. points plotted	1000		

faithfully captures the cantilever dynamics that results when a tip interacts with a substrate. The user can specify the nature of the interaction, the properties of the AFM cantilever, and the feedback response of the AFM controller. A few examples were discussed to illustrate how to simulate a force vs. distance experiment and contact mode AFM topography scans. Future topics covered in Part II of these lecture notes will illustrate dynamic AFM simulations.

The software can be used to advantage when a few parameters are systematically varied to investigate the influence that these parameters have on the outcome of an AFM experiment.

In order to keep track of the input parameters required to plan a sensible simulation of a Force — distance experiment, a convenient worksheet is provided in Table 10.2. In order to keep track of the input parameters required to simulate a contact mode experiment, a similar worksheet is provided in Table 10.3 below.

Table 10.3   A template listing the required input parameters for a contact mode line scan in VEDA.

VEDA Worksheet: Contact Mode Scans

Simulation Notes: _____

Simulation Date:_____

"Operating cond. + Cantilever prop."		"Tip-sample interact. prop.: substrate"	
No. eigenmodes		Tip-sample interaction	
Cantilever stiffness		R_{tip}	
Q		E_{tip}	
Set point deflection		ν_{tip}	
Scan Lines per second		a_o	
P		H	
I		E_{sample}	
Sampling frequency		ν_{sample}	
"Simulation parameters"		"Tip-sample interact. prop.: Feature"	
Num. points plotted	1000	Geometric feature	
		Feature height	
		Length of feature	
		Length of trapezoid top	
		Include geometric convolution	
		Specify material properties	

10.9   Further Reading

References to the Original Literature:

[caveat]   The developers of VEDA are continually upgrading the capabilities of this software. This implies that default conditions for simulations may change over time. Blindly accepting default parameters suggested by VEDA does not guarantee the same default parameters used in the simulations discussed in this book.

[kiracofe12]   D. Kiracofe, J. Melcher & A. Raman, "Gaining insight into the physics of dynamic atomic force microscopy in complex environments using the VEDA simulator" Rev. Sci. Instr. **83**, 013702 (2012).

[melcher08]   J. Melcher, S. Hu, & A. Raman, "VEDA: A web-based virtual environment for dynamic atomic force microscopy" Rev. Sci. Instr. **79**, 061301 (2008).

[nanohub13]   VEDA: Virtual Environment for Dynamic AFM, D. Kiracofe, J. Melcher, A. Raman, S. Balasubramaniam, S.D. Johnson, S. Hu, available at https://nanohub.org/resources/veda. Last accessed March 30, 2014.

[nanohub_intro]   A world-wide resource for nanoscience and nanotechnology, nanoHUB.org was created by the Network for Computational Nanotechnology. The project was funded in large part by the US National Science Foundation. See https://nanohub.org/. Other than establishing a user account, there is no fee associated with the use of the nanoHUB.

[veda_manual]   The VEDA manual can be downloaded as a pdf from https://nanohub.org/resources/veda/supportingdocs.

Further reading:

10.10   Problems

1. Simulate a force vs. distance experiment on samples having a Young's modulus that ranges between 100 GPa to 1 GPa. Assume the Hertz model for tip–sample interaction. Use a cantilever with a 0.5 N/m spring constant. Perform an F-z simulation starting from a z-distance of 5 nm and approach to a z-distance of -2 nm. For the capillary force, choose a pull-off distance of $d = 0.6$ nm. Select an energy dissipation of 1 eV (this is the energy dissipated on withdrawal). Use the default values provided by VEDA for other parameters. The ultimate goal is to compare the resulting indentation for three samples with Young's modulus of 100 GPa, 10 GPa, and 1 GPa. Write a paragraph that quantitatively discusses your results.

2. Perform an approach and retract F-z simulation on a substrate in which the tip–sample interaction is modeled using JKR contact mechanics. First, let z range from $+10$ nm to -5 nm. Then, to capture the backward retraction of the tip, perform a 2nd simulation in which z ranges from -5 nm to $+10$ nm. Zoom into the region near $z = 0$ to better view the hysteretic behavior. Write a paragraph that quantitatively discusses the results of this simulation.

3. You have a very soft sample with an elastic modulus of only 0.5 GPa. It is known that a force of 500 pN or higher will reversibly damage the sample. You have the choice of two cantilevers with stiffness values 0.1 N/m and

1 N/m. Assume the tip–sample contact mechanics is well approximated by the DMT model.

(a) At what z-distances will these two cantilevers jump to contact?

(b) Which cantilever would be the best if you wanted to investigate the relevant tip–sample forces when $d = +0.9\,\text{nm}$?

(c) Using the 1 N/m cantilever, estimate the maximum Z value permissible without causing permanent damage?

For all other parameters not discussed above, use the default values provided by VEDA (use the DMT tip–sample interaction).

Write a paragraph that quantitatively discusses the results of this simulation.

4. Perform a contact mode scanning simulation on a sample that has a trapezoidal height of 10 nm, a top length of 50 nm and a base length of 75 nm.

   (a) Under the Operating conditions tab, set the following parameters to specify the cantilever and the controller properties:

   - One eigenmode for the cantilever,
   - Cantilever stiffness of 3 N/m,
   - Quality factor of 100,
   - Natural cantilever frequency of 40 kHz,
   - Setpoint deflection of 4 nm,
   - Scan lines/second = 10,
   - Proportional and integral gains of 0.005.

   (b) Under the tip–sample interaction properties: substrate tab, choose the DMT contact (with all its default values) to specify the tip–sample interaction.

   (c) Under the Simulation parameters tab, set the scan size to 150 nm. Run a simulation and investigate the mean interaction force as the tip scans across the trapezoidal feature. What do you observe when

   - the tip approaches the feature
   - the tip moves past the feature

Write a paragraph that quantitatively discusses the results of this simulation.

5. Using the Contact mode scanning tool in VEDA, analyze a sample that has rectangular trench geometry 40 nm wide with a depth of -30 nm. The sample has a modulus of 100 GPa. Select the following relevant parameters:

- DMT model
- 1 eigenmode for the cantilever,
- Tip radius 8 nm
- Cantilever stiffness of 0.6 N/m,
- Quality factor of 80
- Natural frequency of 44 kHz
- Setpoint deflection of 2 nm
- Proportional and Integral gains of ~0.001

Investigate the contact mode topography as a function of scan speed. Vary the scan lines/second from 2, 5, 10. Initially, include no tip convolution to make the simulation run faster. After the simulation is working properly, turn tip convolution "ON".

Write a paragraph that quantitatively discusses the results of this simulation.

Index

Printed in the United States
By Bookmasters